nanoelectronic materials laboratory

Technische Universität Dresden

Department of Electrical Engineering and

Information Technology

Dissertation

Formation of Ferroelectricity in Hafnium Oxide Based Thin Films

submitted by

Tony Schenk

born January 11, 1987 in Großenhain, Germany

in partial fulfillment of the requirements for the degree of

Doktoringenieur

(Dr.-Ing.)

Chairman:	Prof. Dr. S. Mannsfeld	Technische Universität Dresden
Supervisor:	Prof. Dr. T. Mikolajick	Technische Universität Dresden
Supervisor:	Prof. Dr. D. Damjanovic	École polytechnique fédérale de Lausanne
4th Member:	Prof. Dr. habil. W.-J. Fischer	Technische Universität Dresden
	date of submission:	2016-09-30
	date of defense:	2016-12-19

Bibliografische Information der Deutschen Nationalbibliothek:
Die Deutsche Nationalbibliothek verzeichnet diese Publikation in der Deutschen Nationalbibliografie,
detaillierte bibliografische Daten sind im Internet über http://dnb.dnb.de abrufbar.

© 2017 Tony Schenk
Herstellung und Verlag:
BoD – Books on Demand, Norderstedt

ISBN: 978-3-7431-2729-6

Abstract

In 2011, Böscke et al. reported the unexpected discovery of ferroelectric properties in hafnia based thin films, which has since initiated many further studies and revitalized research on the topic of ferroelectric memories. In spite of many efforts, the unveiling of the fundamentals behind this surprising discovery has proven rather challenging. In this work, the originally claimed Pca2_1 phase is experimentally proven to be the root of the ferroelectric properties and the nature of this ferroelectricity is classified in the frame of existing concepts of ferroelectric materials. Parameters to stabilize this polar phase are examined from a theoretical and fabrication point of view. With these very basic questions addressed, the application relevant electric field cycling behavior is studied. The results of first-order reversal curves, impedance spectroscopy, scanning transmission electron microscopy and piezoresponse force microscopy significantly advance the understanding of structural mechanisms underlying wake-up, fatigue and the novel phenomenon of split-up/merging of transient current peaks. The impact of field cycling behavior on applications like ferroelectric memories is highlighted and routes to optimize it are derived. These findings help to pave the road for a successful commercialization of hafnia based ferroelectrics.

Autorenreferat

Die erste Veröffentlichung von Böscke et al. über ferroelektrisches Verhalten von Hafniumoxid-Dünnschichten erregte enorme Aufmerksamkeit unter Materialwissenschaftlern und konnte das Interesse an ferroelektrischen Speichern wiederbeleben. Dennoch stellte sich die Erforschung der Materialgrundlagen dieser überraschenden Entdeckung als recht schwierig heraus. In der vorliegenden Dissertation wird die von Böscke et al. vorgeschlagene Pca2_1-Phase experimentell als Ursache der ferroelektrischen Eigenschaften nachgewiesen und eine Einordnung dieser in bestehende Konzepte zur Ferroelektrizität vorgenommen. Geeignete theoretische und prozesstechnische Stellschrauben zur Stabilisierung der Phase werden untersucht. Nach Klärung dieser sehr fundamentalen Fragen wird der Fokus auf das anwendungsrelevante Feld des Zyklenverhaltens der Kondensatorstrukturen gelegt. Die Ergebnisse von "first-order reversal curves", Impedanzspektroskopie, Rastertransmissionselektronenmikroskopie und Piezokraftmikroskopie erweitern das Verständnis der zugrundeliegenden Mechanismen des initialen Konditionierverhaltens, der Polarisationsermüdung sowie des erstmals beschriebenen Effektes der Aufspaltung von Schaltmaxima in transienten Stromkurven. Die Bedeutung des Zyklenverhaltens für Anwendungen wie ferroelektrische Speicher wird ausführlich dargestellt um anschließend Optimierungsansätze abzuleiten. Die Ergebnisse dieser Arbeit tragen dazu bei, den Weg zur erfolgreichen Kommerzialisierung hafniumoxidbasierter Ferroelektrika zu ebnen.

Acknowledgements

I would like to cordially thank the whole **NaMLab staff**. Since I joined the team for my master thesis in April 2012, I have gotten to know a familiar and prolific work climate. Everybody is helpful and willing to consider a problem another person is struggling with. This atmosphere is at least not to be taken for granted and highly appreciable.

I am very grateful to **Uwe Schroeder**, who has supervised me a for the last four years as head of the dielectrics group. He served as a first contact person on at least a weekly base and has given valuable advice how to keep an eye on the big picture. The mixture of guidance on the one hand as well as freedom and trust on the other hand worked very well for me.

My doctoral advisor, **Thomas Mikolajick**, is acknowledged for always having an open door for any scientific and also strategic questions. I always valued the critical, but always open discussions and the trust in me drawing the appropriate conclusions and making the corresponding choices. Moreover, professors who evidently read all publications are, unfortunately, not to be taken for granted.

Particularly, I would like to thank:

- **Johannes Müller** (Fraunhofer IPMS, Dresden), **Ekaterina Yurchuk** and **Stefan Mueller** for paving my way by their work, introducing me to the topic and basic measurement procedures and fruitful exchange especially during the beginning of my work.
- **Claudia Richter** for developing and maintaining the Si:HfO_2 ALD processes and invaluable support in sample preparation as well as **Michael Hoffmann** whose work as a smart, hardworking and ambitious student assistant accelerated basic sample screening and, beyond that, enabled the FORC measurements at NaMLab.
- **Milan Pešić** and **Johannes Ocker** for the many stimulating discussions on electrical characterization and simulation.
- **Felix Schubert** as well as **Steve Knebel** and **Stefan Slesazeck** for quickly being on the scene every time the almighty "D8 Destroyer" or one of the many-legged and -cabled prober monsters endangered my nerves or finishing time.
- **Andreas Krause**, **Hannes Mähne** and **Jan Gärtner** for their introduction to PVD processing and help with tool issues.
- **Helge Wylezich** and **Dominik Martin** for introducing me to impedance analysis and atomic force microscopy, respectively.
- **Ulrich Böttger**, **Sergej Starschich** (RWTH Aachen), **Alfred Kersch**, **Robin Materlik** and **Christopher Künneth** (Munich UAS) for frequent discussions on latest results, conventional ferroelectrics, defect mechanisms and phase stability. It was a pleasure to work with them in our common DFG project.

- **Darius Pohl** and **Bernd Rellinghaus** (Leibniz IFW, Dresden) as well as **Everett D. Grimley, Xiahan Sang** and **James M. LeBeau** (North Carolina State University, Raleigh) for their contribution to the TEM studies in this work. A lot of fruitful discussions advanced my understanding of TEM and helped developing suitable experiments.
- **Jörg Grenzer** (Helmholtz-Zentrum Dresden-Rossendorf), **Hiroshi Funakubo** (Tokyo Institute of Technology), **Jacob L. Jones, Christopher M. Fancher** (North Carolina State University, Raleigh) and **Min Hyuk Park** for all the fruitful discussions about X-ray diffraction and the crystallography of the hafnia phases. To Chris, additional thanks for his support with the Rietveld refinement.

Furthermore, I would like to thank the German Research Foundation (**Deutsche Forschungsgemeinschaft, DFG**) for funding my work within the project "Infercx" (MI 1247/11-1). The scanning probe measurements described in section 5.5 of this work were conducted within the CNMS User Program 2015 at Oak Ridge National Laboratories. The **Center for Nanophase Materials Sciences**, which is a DOE Office of Science User Facility, is acknowledged for funding this part of my research. A special thanks goes to **Nina Balke** (Wisinger), **Sergei V. Kalinin, Stephen Jesse, Anton V. Ievlev, Evgheni Strelcov, M. Baris Okatan** and **Alexei Belianinov** for their technical and personal support.

Last but not least, **my family** and all **my friends** are acknowledged. Luckily, my hubristic and outstandingly cocky character—one of my rare talents—prevented any significant personal crises and doubts. Therefore, I would rather like to thank them for their support and for enduring my repeated excuses containing any of the words "diss", "paper", "conference",…, "deadline". **I hope, all beers I missed were carefully kept track of! I am more than willing to arrange a payback with all of you!**

Index

Index	I
1 Motivation	1
2 Fundamentals of Ferroelectricity and Ferroelectric Memories	3
2.1 Ferroelectricity from the Electrical and Structural Point of View	3
2.2 Phenomenological Theory of Ferroelectricity and Phase Transitions	7
2.3 Preisach Model of Ferroelectric Hystereses	11
2.4 Assessment of Non-ideal Ferroelectric Hystereses	12
2.5 Non-volatile Random Access Memory and Ferroelectric Memory	22
2.6 Hafnium Oxide for Ferroelectric Memories	28
3 Sample Preparation and Characterization Methods	31
3.1 Fabrication of Metal-Insulator-Metal Capacitors	31
3.2 Basic Methods	33
3.3 Scanning Probe Microscopy	36
3.4 Transmission Electron Microscopy	42
3.5 Impedance Spectroscopy	46
3.6 First-Order Reversal Curves (FORC)	50
3.7 Harmonic Analysis	52
4 Stabilization of the Ferroelectric Phase in Hafnium Oxide	55
4.1 Previous Work and Simulation Results	55
4.2 Proof of the Ferroelectric Phase	62
4.3 Impact of Smaller and Larger Dopants	65
4.4 Influence of Film Thickness and Annealing Conditions	76
4.5 Electrode Effects	79
4.6 Texture of the Ferroelectric Thin Films	82
4.7 HfO_2—An Incipient Ferroelectric?	87
4.8 Summary	89
5 Electric Field Cycling Behavior of Ferroelectric Capacitors	91
5.1 Wake-up Effect	92
5.2 Polarization Fatigue	96
5.3 Split-up and Merging of Switching Peaks in Transient Currents	99
5.4 Discussion of Activation Energies	104
5.5 PFM to Study Switching on Nanoscale	109
5.6 Structural Evolution During Field Cycling	117
5.7 Routes to Improve the Field Cycling Behavior	127
5.8 Summary	128

6 Summary and Conclusion 130

Bibliography XII

A Appendix: Dopant Overview from J. Müller's Dissertation XXXVII

Figures XLI

Tables LI

Abbreviations LIII

Symbols LVIII

Curriculum Vitae LX

List of Scientific Publications LXIII

1 Motivation

The present dissertation has two major objectives:

1. providing a fundamental understanding of the origin of the recently discovered ferroelectric behavior[1] in HfO_2 based thin films, and
2. elucidating the non-ideal, evolution of the macroscopic ferroelectric properties of these films as a result of cyclic switching as it is done in ferroelectric memories and other applications.

Despite its material related character, this thesis has to be grasped in view of the bigger picture composed of two important trends in semiconductor industry: On the one hand, there is a shift from computation-centric to data-centric applications driven by e.g. growing numbers of cloud services and large data centers for services like Google, Facebook, Twitter and many others, which all demand immediate availability of data.[2, 3] On the other hand, mobile devices such as smart phones and tablet computers are gaining increasing market shares exceeding those of personal computers (desktop and notebook).[4, 5] These technology branches are important parts of a vision called "The Internet of Things"[6, 7, 8, 9]. Besides the aspiration for continuous miniaturization (Moore's law[10]) and its related technological challenges, additional aspects need to be addressed: Energy consumption and closing the memory gap. The former is easily anticipated knowing that 3 – 7 % of global electrical energy is consumed by computing devices[11] and that about 50 % of the electrical energy consumption of data centers goes into cooling[12, 13]. With huge amounts of data to be kept at prompt access, the memory gap—the access time difference between non-volatile mass storage and volatile random access memory—becomes more and more critical. Today's mainstream memories already struggle to fulfill the current requirements and therefore, research of potential future solutions is in great demand.

Ferroelectric memories with proven low access times, unlimited cycling endurance and non-volatile storage even at elevated temperatures are already on the market.[14, 15, 16, 17] However, their use is limited to niche applications as their current scaling node of 90 nm lags far behind of those of flash memory or DRAM.[18, 14] These limitations are due to issues related to conventional ferroelectric materials and right here, HfO_2 comes into play. Hafnia is a state-of-the-art high-k dielectric[19] with mature processing technologies. Therefore, the discovery of its ferroelectric behavior in 2007 at the memory company Qimonda and first publication by Böscke et al.[1] in 2011 revived the interest in ferroelectric memories as contender for future memories.

However, on the way from the discovery of an effect to a successful commercialization, there are important intermediate steps to be taken: Firstly, a fundamental material understanding eases finding the right knobs to obtain the desired material properties. Secondly, stabilizing these properties in lab-scale structures as well as an understanding of perturbing effects and how to mitigate them is needed. Thirdly, an integration scheme into state-of-the-art technology has to be developed. Finally, statistically reliable operation of these devices under a variety of environmental conditions as required by potential customers has to be demonstrated. The objectives of this thesis listed at the beginning specifically target the first two points and aim to pave the way toward a successful commercialization of hafnia/zirconia based ferroelectrics in ferroelectric memories. As ferroelectricity is a rather noble material property implying also useful electrothermal and electromechanical properties, a lot of other future applications will hopefully profit from this work.

2 Fundamentals of Ferroelectricity and Ferroelectric Memories

2.1 Ferroelectricity from the Electrical and Structural Point of View

"Ferroelectricity" describes a non-linear dielectric behavior. Although it contains the word "ferro" (lat. iron), it is not related to any properties of the chemical element Fe, but was purely adapted by analogy to ferromagnetism. Ferromagnetism describes the hysteretic relation of the applied magnetic field (strength) H and the resulting magnetic flux density B. The corresponding physical quantities of ferroelectricity are the electric field (strength) E and the (electric) displacement field D.[20]

In a simple planar and homogeneous capacitor arrangement, the displacement field represents an areal charge density (charge Q per area A) caused by an external electric field applied to a material with the (dielectric) permittivity ε:

$$\frac{Q}{A} = D = \varepsilon E = \varepsilon_0 \varepsilon_r E. \tag{2.1}$$

This dielectric permittivity ε is usually expressed as the product of electric field constant ε_0 [1] and the relative permittivity ε_r (see. Eq. (2.1)). Following this definition, the latter expresses the multiplication of the vacuum permittivity due to the presence of the material under investigation. In semiconductor industry, ε_r is commonly referred to as k or the k-value of a dielectric. Moreover, it is common to separate the material contributions, the (dielectric) susceptibility χ, from the vacuum contribution:

$$\varepsilon_r = \varepsilon_0 + \varepsilon_0 \chi. \tag{2.2}$$

Similarly, the displacement field D is, thus, split into the displacement field of vacuum D_0 and the dielectric polarization density P—for the sake of brevity usually only referred to as polarization:

$$D = D_0 + P = \varepsilon_0 E + \varepsilon_0 \chi E. \tag{2.3}$$

For $\chi \gg 1$,

$$D \approx P. \tag{2.4}$$

However, instead of plotting D vs. E in analogy to what is done for ferromagnets, a graph of P vs. E is applied as shown in Fig. 2.1. Because of the high permittivity values of conventional ferroelectrics, which are in the order of several hundreds, the vacuum contribution

---
[1] also known as vacuum permittivity

is negligible. Moreover, at zero field the difference between P and D vanishes. The two y-intercepts represent the positive and negative remanent polarization P_{r+} and P_{r-}, respectively. The term "remanent" refers to the fact that these are the two polarization quantities that remain when the external electric field is removed. The x-intercepts are called positive coercive field E_{c+} and negative coercive field E_{c-}. "Coercive" points to the fact that these are the respective E values necessary to cancel the remanent polarization by an external field to achieve an effective polarization of 0 again.

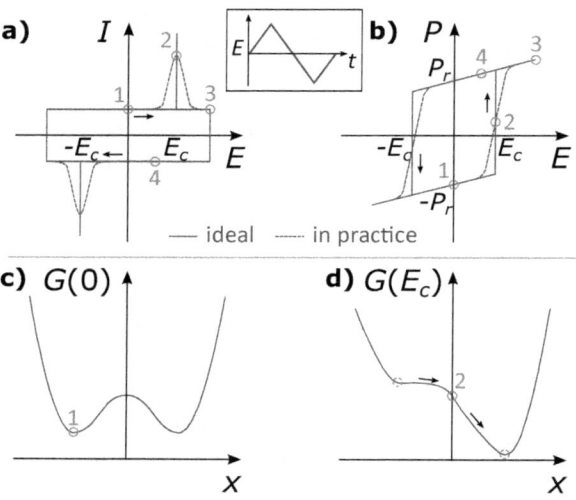

Fig. 2.1: a) Transient current response of b) a polarization hysteresis measurement via triangular field sweeps (inset). The corresponding atomic double-well potential at c) zero field and d) around the coercive field are shown at the bottom.

Such a P-E curve can be obtained by applying a triangular field sweep of an amplitude high enough to fully switch the ferroelectric. Fig. 2.1 shows the evolution of the measured transient current I, which is integrated over time to obtain the charge in Eq. (2.1) and, with it, the polarization P. In an ideal case, $I(E)$ consists of a constant displacement current proportional to the capacitance C ($I \propto C \cdot dE/dt$) of the sample and additional switching peaks around E_c where the polarization is inverted. The branches of the P-E curve outside the hysteretic region are referred to as positive and negative saturation regions. Deviations from this very ideal scenario are discussed in section 2.4.

The observed polarization hysteresis exhibits two stable states, P_{r+} and P_{r-}, which remain after the external applied field is removed. At the atomic scale, they correspond to two stable positions of one or more ions in a crystallographic unit cell. A stable position is defined by an energetic minimum. Thus, a bistable behavior originates from a double-well shaped curve of Gibbs energy G dependent on the ion position (less precisely often abbreviated as: double-well potential) as it is shown in (Fig. 2.1). In contrast, this potential has to equal a harmonic

2.1 Ferroelectricity from the Electrical and Structural Point of View

oscillator potential, i.e. a parabola shape, in order to achieve what is called "linear dielectric behavior":

$$G \propto x^2. \tag{2.5}$$

Some simple derivations for this fundamental model case will ease the understanding of what will be discussed in the following sections. The first derivative of the energy G with respect to position x equals the force necessary to displace the ion by distance x from its stable position (minimum of G) at $x = 0$.

$$F = \frac{dG}{dx} \propto x \propto F_{Coulomb} = Q_{ion} \cdot E \tag{2.6}$$

In electrodynamics, this force is given by the Coulomb force $F_{Coulomb}$. For a given ionic charge Q_{ion}, x is directly proportional to the applied electric field. This displacement x gives rise to an induced dipole moment $p = Q_{ion} \cdot x$ (Eq. (2.6)). The macroscopic polarization P (or displacement field D) as the sum of all these induced dipoles of all unit cells is, thus, also directly proportional to x or E, which is the definition of linear dielectric behavior (Eq. (2.7)).

$$P = \sum p = \sum (Q_{ion} \cdot x) \propto E \tag{2.7}$$

Only in this special case, the relative permittivity k is independent of the applied electric field. A look at the sketched atomic potential in Fig. 2.1 shows that linear dielectric behavior, at best, is to be expected for fields far away from the switching conditions. In section 2.2, a phenomenological model for the shape and temperature dependence of the double-well potential and its consequences for the dielectric behavior of a ferroelectric materials are discussed.

Up to this point, the phenomenon of ferroelectricity has been described in analogy to ferromagnetism and its macroscopic electric manifestations have been linked to the atomic potential. The remaining question concerns the identification of potential ferroelectric materials: Which crystallographic requirements have to be met by a material to exhibit ferroelectricity?

Compared to the case of a linear dielectric, a ferroelectric exhibits net charge that is displaced from the center to one of the two possible eccentric positions. This results in a spontaneous polarization of the unit cell—a polarization is already present without an applied field. Therefore, the crystal has to be non-centrosymmetric, which presents a first necessary but not a sufficient requirement for ferroelectricity. Non-centrosymmetry is a prerequisite for other closely related phenomena, namely piezo- and pyroelectricity. The piezoelectric effect describes a polarization induced by mechanical strain (or the other way around, which is then called inverse piezoelectric effect). In a centrosymmetric crystal, the same amount of charge would be displaced in opposite directions, so no net polarization occurs. Crystallography distinguishes 32 so-called crystal classes. 11 of them are centrosymmetric and can thus not be piezoelectric. Moreover, class 432 is also not able to exhibit piezoelectric behavior because of its high rotational symmetry. As a result, only 20 crystal classes are piezoelectric. Ten of these 20 classes already possess a dipole without being strained—they are polar. As already mentioned, the atomic potentials are temperature dependent and thus, it is not surprising that also the degree of polarity (or a bit simpler: the eccentricity of ions) is a function of temperature. The pyroelectric effect described the polarization change induced by a temperature change. Every ferroelectric is therefore also a pyroelectric and every pyroelectric is also a piezoelectric. Fig. 2.2 shows this subgroup-relation. The only difference of a pyro- and a ferroelectric is the fact that the spontaneous polarization of a ferroelectric can be inverted by an electric field. No further crystallographic requirements are involved and thus, there is no real fundamental way to distinguish these materials. If a material is subjected to a higher and higher field, it either inverts its polarization or suffers dielectric breakdown. The latter only justifies the statement that the field to switch it—if this is possible—is higher than the breakdown field. The performed experiment was just not able to move the ion to the second potential minimum. From a crystallographic point of view, nothing speaks against this possibility: If the unit cell is simply mirrored in the direction of the field, it is still the same crystal class with inverted polarity (as favored by the external field) and atomic potentials. Consequently, the differentiation of pyro- and ferroelectrics is just an empiric one.[21]

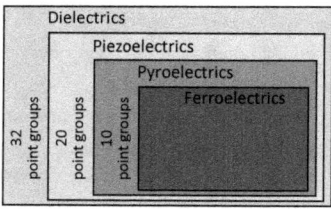

Fig. 2.2: Classification of ferroelectric within the group of (crystalline) dielectrics: Ferroelectric are a subgroup of pyroelectric, which themselves are a subgroup of piezoelectrics.

Fig. 2.2 explicitly refers only to crystalline, dielectric materials. For metals, an external field would not be able to displace any atoms or ion cores, because it is immediately compensated by current flow as the result of a flux of free electrons. All considerations of electric quantities as a result of or cause for changes in thermal or mechanic properties are, thus, insignificant for this work. Moreover, most metals posses highly symmetric cubic lattices, which exclude them from further consideration from another point of view. There is also a number of ferroelectric polymers including the most common co-polymer of polyvinylidene fluoride with tetrafluoroethylene (PVDF:TrFE).[22, 23] These polymers are also not considered in this section as they are not relevant for this work.

One of the most common conventional ferroelectrics existing in perovskite structure is lead-zirconate-titanate (Pb(Zr,Ti)O_3, PZT). Pure PbZrO_3 exhibits a rhombohedral crystal structure, whereas pure PbTiO$_3$ is tetragonal. Around a Zr:Ti ratio of 52:48, a morphotropic (morphotropy: change of crystallographic structure with composition[24, 25]) phase boundary is found. Piezoelectric properties and k-value become maximized. However, for ferroelectric memories, these properties are not of utmost interest. Rather the opposite is true regarding the k-value as will be shown in section 2.5: If the permittivity is very high, this is undesired because it reduces the ratio of switching to non-switching charge and increases the field drop across interfacial layers. Hence, Zr:Ti ratios of 0.3:0.7 to 0.4:0.6 are commonly used.[26, 27] Fig. 2.3 illustrates the two stable positions of the central Ti or Zr ion in the center of the tetragonal Ti-rich PZT cell, which represent the two polarization states of the ferroelectric material (compare Fig. 2.1).

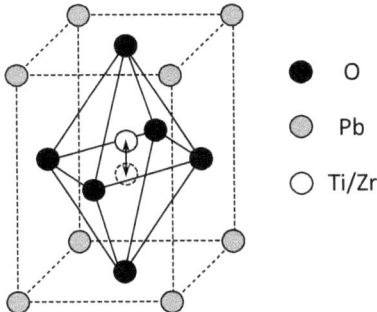

Fig. 2.3: Sketch of the tetragonal perovskite unit cell of PZT.

The temperature dependence of the ferroelectric properties are subject of the following section.

2.2 Phenomenological Theory of Ferroelectricity and Phase Transitions

As discussed in section 2.1, ferroelectric properties originate from an atomic double-well potential. With increasing temperature, these properties vanish and two possible explanations for the phase transition exist[22, 28]:

1. **Order-disorder phase transition**: As thermal energy ($k_B T$) reaches the value of the energy barrier between the minima in the double-well potential, more and more ions freely are able to statistically switch between the two stable states. The statistical average occupation of both states becomes 0.5 and thus, the observed net polarization becomes zero. The double-well potential is assumed to be nearly temperature independent. Another explanation is that the long-range-order of the dipols and thus, the macroscopic net polarization gets lost.

2. **Displacive phase transition**: The temperature dependence of the double-well potential dominates and the vanishing barrier is the major effect compared to the increase in thermal energy.

Thermal energy at room temperature is about 26 meV. As will be shown later in section 4.1, the energy barrier between the two polarization states or between a ferroelectric and a paraelectric state are at least 35 meV for hafnia and zirconia. Thus, this section is devoted to the phenomenologic theory of a displacive phase transition.[28]

A phenomenological approach to describe the dielectric behavior of ferroelectric near its transition temperature from a ferroelectric into a paraelectric phase is based on the theory by Landau and Devonshire [29, 30, 28]. The basis of this theory is a series expansion of Gibbs energy as a function of polarization P as the order parameter[28]:

$$g = \frac{G}{V} = \frac{1}{2}c_2 \cdot P^2 + \frac{1}{4}c_4 \cdot P^4 + \frac{1}{6}c_6 \cdot P^6 - E \cdot P, \tag{2.8}$$

with c_2, c_4 and c_6 being constants. All powers higher than 6 are truncated and for reasons of symmetry between the two polarization states, only even powers of P are included. Considering the explanation in section 2.1 and especially Eqs. (2.5) to (2.7), this power series expression also represents the shape of the atomic potential for the switching ion(s). Instead of a harmonic oscillator, a sixth-order polynomial is applied to model the dielectric behavior. In addition to that series expansion, the term $-E \cdot P$ accounts for the contribution of an external electric field. It can be seen, that a positive field reduces the Gibbs energy for a positive polarization whereas a negative field favors a negative polarization. A stable state equals a minimum in G, or Gibbs energy density g as in the notation used below, and can thus be obtained by the criterion $dG/dP = 0$, which provides an expression for the electric field E as a function of P:

$$\frac{dg}{dP} = 0 \rightarrow E = c_2 \cdot P + c_4 \cdot P^3 + c_6 \cdot P^5. \tag{2.9}$$

The sufficient condition to make this extremum a minimum is

$$\frac{d^2 g}{dP^2} = (\varepsilon_0 \chi)^{-1} = c_2 + 3c_4 \cdot P^2 + 4c_6 \cdot P^4 > 0. \tag{2.10}$$

Above the transition temperature, P becomes zero and the susceptibility is given by:

$$\varepsilon_0 \chi = \frac{1}{c_2}. \tag{2.11}$$

The temperature dependence of the dielectric behavior around the transition temperature results from the assumption that this factor c_2 depends linearly on the temperature, which directly equals the expression of the Curie-Weiss law:

$$c_2(T) = c_2' \cdot (T - T_0) = (\varepsilon_0 \chi(T))^{-1} \tag{2.12}$$

2.2 Phenomenological Theory of Ferroelectricity and Phase Transitions

Including the temperature dependence of c_2 in Eq. (2.8) gives a general expression for the shape of the atomic potential at different temperatures T.

$$g = \frac{G}{V} = \frac{1}{2}c_2' \cdot (T - T_0) \cdot P^2 + \frac{1}{4}c_4 \cdot P^4 + \frac{1}{6}c_6 \cdot P^6 - E \cdot P \qquad (2.13)$$

Eq. (2.13) marks the starting point for the discrimination of two different types of phase transitions to be discussed in the following. Depending on the coefficients c_2, c_4 and c_6, a phase transition of first (discontinuous) and second (continuous) order are distinguished. First of all, c_2 has to be positive to ensure a positive susceptibility of the non-polar state as it is evident from Eq. (2.11). Secondly, the coefficient of the highest order-term has to be always positive. Else, g would become minimal for $P \to \pm \infty$. With P, also x approaches $\pm \infty$, which corresponds to the decomposition of the crystal. Two cases are distinguished[21]:

- $c_6 \approx 0$ and $c_4 > 0$ → second-order transition
- $c_6 > 0$ and $c_4 < 0$ → first-order transition

This choice of c_4 and c_6 affects the temperature dependence of both susceptibility χ and spontaneous polarization P_s. Since the **second-order transition** is the easier case, it is discussed first. The starting point for derivations regarding F_s is equation Eq. (2.9). At zero field and with $c_6 \approx 0$, two options to solve the equation can be found:

$$0 = c_2 \cdot P_s + c_4 \cdot P_s^3 = c_2' \cdot (T - T_0)P_s + c_4 \cdot P_s^3 \qquad (2.14)$$

$$0 = \begin{cases} P_s & \to \quad \text{trivial solution} \\ c_2' \cdot (T - T_0) + c_4 \cdot P_s^2 & \to \quad \text{non-trivial solution.} \end{cases} \qquad (2.15)$$

From the non-trivial solution in Eq. (2.15), a square-root-shaped, continuous decay of P_s as shown in Fig.2.4 follows. In a perfectly oriented single crystal, P_s equals the macroscopically measured P_r, for $T < T_0$.

$$P_s = \pm \sqrt{\frac{c_2' \cdot (T_0 - T)}{c_4}} \qquad (2.16)$$

The phase transition temperature T_0 in Eq. (2.16) defines what is called the Curie temperature T_C. Inserting this solution for P_s into Eq. (2.10), the temperature dependence of χ^{-1} below T_0 can be obtained. Below T_0, the slope of $\chi(T)^{-1}$ is precisely double the value as above T_0, but has a negative sign. For $T = T_0$, χ approaches infinity

$$(\varepsilon_0 \chi(T))^{-1} = 2c_2' \cdot (T_0 - T) \qquad (2.17)$$

Fig. 2.4 summarizes the temperature dependent evolution of dielectric properties for the second-order transition.

Similar derivations are possible for the **first-order transition**. Due to the additional sixth-order term, the results below T_0 differ:

$$P_s^2 = \frac{1}{2c_6}\left(|c_4|\sqrt{g_4^2 - 4c_2'(T-T_C)g_6}\right) \qquad (2.18)$$

$$(\varepsilon_0 \chi(T))^{-1} = \frac{3g_4^2}{4g_6} + 8c_2' \cdot (T_0 - T) \qquad (2.19)$$

The main difference, however, is that T_C and T_0 do not coincide. At T_C, the polar phase exhibits exactly the same value of G as the non-polar phase, as can bee seen from Fig. 2.5. The existence of metastable states between T_C and T_0 results in a discontinuity between the branches with ascending and descending $\chi(T)$. In practice a hysteresis occurs when passing the transition upon heating or cooling and the jump might be found at different temperatures depending on when the metastable state relaxes into the new stable state. For reasons of completeness, it should be mentioned, that further characteristic temperatures above T_C can be introduced: One describes the point, where the polar phase loses its metastability, i.e. the eccentric local minima at zero field vanish and at even higher temperatures, there is a second temperature, at which the polar phase cannot be stabilized even under applied field.[31]

Fig. 2.5 gives an overview about the dielectric properties at different temperatures for a phase transition of first order. In the transition region, this potential enables the observation of a field-induced phase transition if the applied electric field is high enough. For increasing E, the term $-E \cdot P$ tilts the potential and reduces the barrier between the centric well and the eccentric minima.

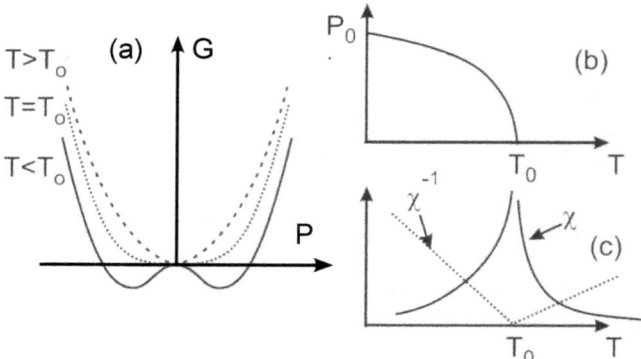

Fig. 2.4: Second-order transition: (a) Free energy as a function of polarization (ion position) at different temperatures, (b) spontaneous polarization (here, P_0) and (c) susceptibility as function of temperature.[28]

2.3 Preisach Model of Ferroelectric Hystereses

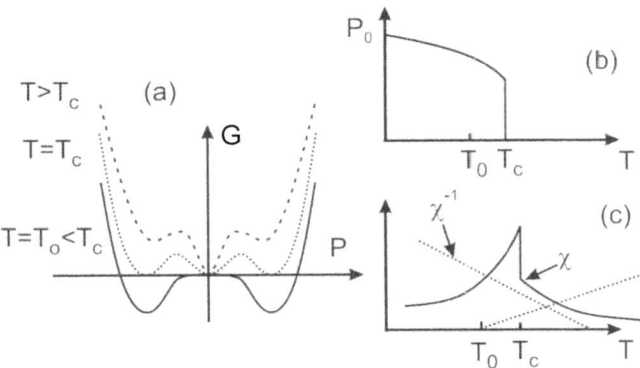

Fig. 2.5: First-order transition: (a) Free energy as a function of polarization (ion position) at different temperatures, (b) spontaneous polarization (here, P_0) and (c) susceptibility as function of temperature.[28]

Concluding this section, it should be mentioned that this phenomenological description and the differentiation between a first- and second-order transition is based on the assumption of the atomic potential exhibiting a certain polynomial shape as shown in Eq. (2.8). Similar to the harmonic oscillator model discussed in section 2.1, it helps to understand to observations in reality by simplification and idealization. Deviations from these models in reality are not scenarios to be called "unusual" and are rather cases which, by chance, do not fulfill the arbitrarily chosen prerequisites of the model.

2.3 Preisach Model of Ferroelectric Hystereses

The Preisach model [32] describes a macroscopic hysteresis as the sum of contributions by microscopic bistable units (sometimes called "hysterons"[33]).[34] Such a bistable unit can be conceived as a single domain or in the most nanoscopic case represent one unit cell. Each elemental hysteresis possesses a certain field for switching to the positive polarization state $+p$ (switching field E_s) and a certain field for switching to the negative polarization state $-p$ (backswitching field E_{bs}). In analogy to an ideal rectangular macroscopic P-E hysteresis, the hysterons can be understood as p-E hysteresis of different width ($2 \cdot E_c$ and, to account for asymmetry, an offset towards positive or negative fields (bias field E_{bias}).

These assumptions imply an inhomogeneous behavior of the sample and thus, are to greater or lesser extent in conflict with the model approaches presented in the previous section 2.2. The (back-)switching field, which is linked to overcoming the potential barrier between the polarization states, is allowed to differ between the hysterons. This field originates from the elemental double-well potentials as the base of the temperature dependent changes in dielectric properties. Therefore, the microscopic contributions to χ or P_s are anticipated to differ. The higher the scatter in E_s and E_{bs}, the more likely the deviations from what was discussed in section 2.2. Moreover, even the symmetry with respect to E is no longer a premise. This is opposed to Landau's symmetry arguments and thus, hints on extrinsic contributions or non-ideal unit cells.

Real ferroelectric samples always contain defects and these defects account for non-ideal behavior. To improve the performance of a ferroelectric used in a certain application, it is important to understand the nature of the defects to later on choose the right knobs to reduce them or render them inert. An assessment of the distribution of the switching fields, i.e. the switching density or also referred to as "Preisach distribution/density", is helpful. Section 3.6 explains how such a Preisach distribution can be measured. In the following section 2.4, the focus is on how imperfections manifest in the macroscopically measured P-E hysteresis. The section concludes describing how switching density plots can help to distinguish between different sources for non-ideal hysteresis loops.

2.4 Assessment of Non-ideal Ferroelectric Hystereses

The hysteretic curve of polarization vs. electric field is a central property of a ferroelectric. The axis intercepts, P_r and E_c are important characteristics for applications such as ferroelectric memories as will be discuss in section 2.5. For this application, the stability of a polarization state (data retention) as well as the ability to withstand a highest possible number of switching cycles (cycling endurance) are of crucial interest. Deviations from a non-ideal hysteresis shape or changes of the shape during the course of cyclic operation have to be understood to choose the right measures to mitigate the underlying mechanisms. In practice all hystereses are to a certain degree non-ideal. Especially for the rather new ferroelectrics based on HfO_2, hystereses often look rather rounded and for a pristine sample even pinched and asymmetric. To properly assess P_r and E_c, a preconditioning treatment consisting of a certain number of switching cycles was commonly applied.[35, 36, 37] Pinching and asymmetry vanished. Similar observations exist for conventional ferroelectrics such as PZT. The original hysteresis shape was referred to as "constricted"[38, 39, 40, 41], "pinched"[38, 40], "aged"[38, 42, 43, 41], "not fully open"[38, 40], "propeller-shaped"[44, 45] or "double-loop"[46, 47] in contrast to the case after cycling, which was described as "non-constricted"[38], "depinched"[38, 40], "deaged"[38, 42, 43], "open"[38, 39], "square-shaped"[44] or "relaxed"[40, 41].

In literature, a lot has been done to study the manifestation of imperfections on the behavior of ferroelectric materials. However, when studying a material for e.g. a memory application, an electrical measurement is more easily at hand than an in-depth structural study. And for these sophisticated structural studies, an indicator on where to start is always desirable to concentrate the efforts into the most promising directions. So here, the question is precisely the opposite: What information can be taken from a measured P-E hysteresis to derive conclusions about underlying imperfections? This question motivated a recent review article

2.4 Assessment of Non-ideal Ferroelectric Hystereses

on the topic.[48] A summary regarding the assessment of P-E hystereses is given in the following.

First of all, it is necessary to recall the difference between the ideal and the real ferroelectric P-E and transient I-E curves shown in Fig. 2.1. For the easiest case of triangular field sweeps, the transient currents are constant with a sign equaling the sign of the constant slope of $E(t)$. An ideal ferroelectric would switch and backswitch at a certain field E_s and E_{bs}, respectively, which would manifest itself in infinitely sharp and high Dirac peaks in the current at these fields. This would result in an infinitely steep slope of the hysteresis between to two non-switching branched and the switching fields would perfectly coincide with the coercive fields. In practice, inhomogeneities cause a certain distribution of the switching fields and the peaks become rounded as does the hysteresis curve.

To model different phenomena and their manifestation, the ideal linear dielectric contribution was left untouched, i.e. the rectangular-shaped underground in the I-E curve remains. To model the distribution of switching fields, different Gaussian distributions were applied around different fields and added to the linear contribution before numerically integrating the resulting I-E curve to obtain the corresponding P-E hysteresis. Fig. 2.6 summarizes the respective curves for different scenarios explained below.

14   2 Fundamentals of Ferroelectricity and Ferroelectric Memories

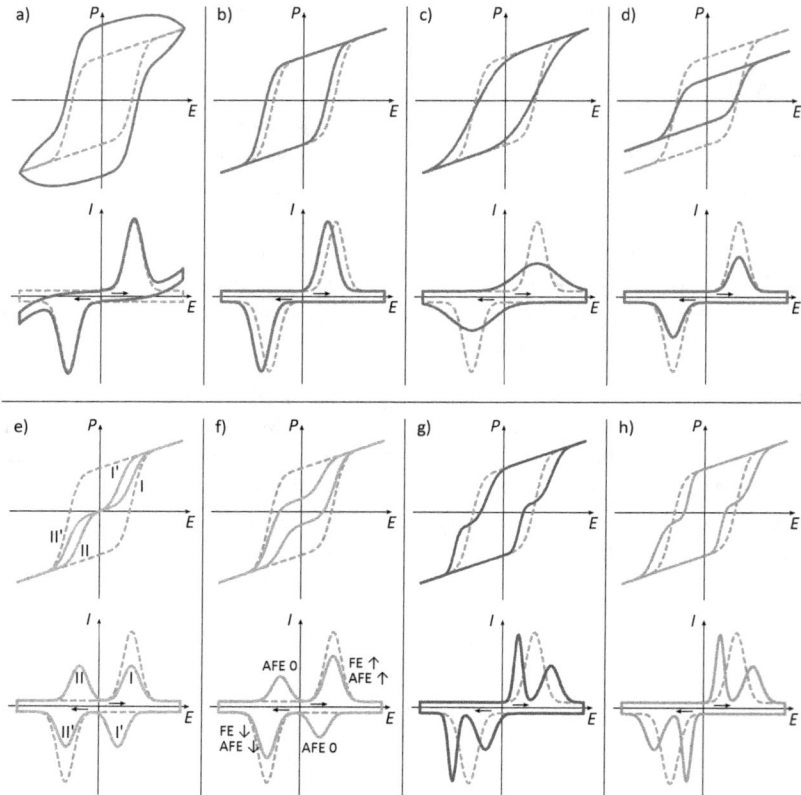

Fig. 2.6: Different scenarios with polarization hystereses (*P-E* plots) in the upper and the corresponding current response to a triangular excitation signal (*I-E* plots) in the lower part of a) – h), respectively. Explanations for possible causes of the curves are given in the text.[48]

a) **Leakage current** contributions can cause an inflated hysteresis as shown in Fig. 2.6 a) top. The leakage current is modeled as a simple arbitrary exponential function of the absolute electric field to mimic the dependence observed in practice. Therefore, the current level beyond the switching fields does not settle to the value given by the linear dielectric component. The exponential increase of the leakage current dominates. During the subsequent field ramp of negative slope, this results in an effective current that is still positive instead of dropping immediately to a constant negative value. I.e., although the field is swept downward, the polarization still increases artificially because of the integrated positive current. In an extreme case, where leakage current is orders of magnitude higher than the ferro- or dielectric components, a circular curve is observed. If the current compliance of the measurement system is reached, the current is cut to

2.4 Assessment of Non-ideal Ferroelectric Hystereses

a constant value at a certain field. As soon as this happens, the polarization increases in linear branches. If this compliance is already reached around zero field, a rhombus with the corners at the E and P axis is observed.

The influence of leakage current has already been described by Lines and Glass in the 1970s[31]. However, 30 years later, Scott critically pointed out that many authors had still not considered this artifact when interpreting their results as of ferroelectric origin. In "Ferroelectrics go bananas"[49], he demonstrates P-E measurements on the "leaky" skin of a banana, which result in curve shapes similar to what had been proposed to evidence ferroelectricity by other authors in the passed years: Cigar-shaped curves without concave regions as expected around the switching fields. He also referenced further publications on similar spurious artifacts, that are beyond the scope of this dissertation.

The so-called "dynamic leakage current compensation" (DLCC)[50, 51] allows to subtract the leakage contributions. The method is based on hysteresis measurements on two adjacent frequencies, but only works if 1) the leakage is not orders of magnitude higher than the FE and dielectric components, 2) the switching fields do not strongly differ because of a strong frequency dependence of E_c or 3) the hysteresis shape does not change during repeated switching. In an early process development stage, this might be sufficient to identify FE properties. As soon as application relevant studies are intended, process optimization is necessary to reduce leakage currents. On the contrary, if the produced samples are too leaky to extract a proper hysteresis even with DLCC, process improvement is to be prioritized over applying sophisticated electrical or structural studies. It is evident that the material is rich in defects, that might hamper a clean interpretation of the anticipated outcome of these methods. Moreover, defects affect the phase stability (see section 4.1) and at some point destabilize the desired FE phase. Consequently, there is an increased risk of wasting time and resources studying such samples.

b) If the P-E hysteresis appears displaced towards positive or negative fields (Fig. 2.6 b) top), this is called "**imprint**"[22] . An "internal bias field" or "built-in field"[22] shifts the switching peaks in the current characteristic toward one side of the field. If the shift is strong, i.e. the bias is in the range of the amplitude of the AC field sweep, one side of the hysteresis cannot be saturated, which can give rise to a decrease in the measured P_r. These shifts are defined with respect to E_{c+} and E_{c-} and stem from internal or external (not related to the FE layer itself) origins. Space charge layers formed by an oxygen deficient interface[52], a bulk screening model[53], and the alignment of defect dipoles towards a present domain structure[41, 22] were discussed in literature. Moreover, an asymmetry can be due to different work functions of different electrodes or differently processed electrodes of the same material. The work function difference drops across the FE layer when the electrodes are shorted and has to be compensated by the external field applied during the hysteresis measurement. Therefore, it is recommended to rule out such effects during basic material characterization by choosing a stack that is as symmetric as possible.[22]

c) A look at Fig. 2.6 c) reveals two main differences to the dashed reference curves: 1) The switching peaks are broader, which results in strongly rounded and slanted polarization hysteresis. 2) Part of the positive and negative current peak is moved into the second and fourth quadrant, respectively, which reduces P_r. This situation is frequently observed for polycrystalline thin films. A certain distribution of grain sizes always exists. In bulk ceramics with grain diameters of more than 1 µm, this distribution is barely recognizable in the hysteresis. As soon as the grain sizes are lower than about 100 nm (to give a rough figure), surface energy no longer plays a negligible role in terms of phase stability. The thinner the films, the higher the impact of **grain size variation** on the atomic potentials, which is the origin of the local E_c values as explained in section 2.2. Thus, a pure broadening can be explained by these grain size effects.

However, they do not explain a movement of part of the peak toward the opposite side of the P axis. As explained by Robels et al.[54], this feature in Fig. 2.6 c) can be mathematically modeled by a purely dielectric layer in series to the ferroelectric. A quite comprehensive picture exists in literature. Ab-initio calculations by Stengel and Spaldin[55] demonstrated the occurrence of such a **"dead layer"** or **"passive layer"** for epitaxial $SrTiO_3$. An interface layer of lower permittivity was observed, which significantly lowered the field drop across the FE film. The higher the thickness and the lower the k-value of such a layer, the higher the field drop across it (capacitive voltage divider). Fig. 2.6 c) only shows qualitative changes, because a comparison of one and the same film with and without passive series layer is hardly feasible. For a better understanding of the impact of such interface layers, Eq. (2.21) can be used. For simplicity, the two interfaces are treated as one linear dielectric layer. Moreover, it is assumed that the curve $P_{FE}(E_{FE})$ of the purely FE layer is known. The resulting curve $P(E)$ caused by an additional interface layer of arbitrary permittivity and thickness can be calculated for any applied field $E = V/d$ for a voltage V across the whole film stack of thickness d. The approach of charge conservation in a series capacitor arrangement (while still ignoring D_0 in the FE as argued in section 2.1) is used. The indices FE and IF represent the ferroelectric and the linear dielectric interface layer, respectively. From charge conservation

$$P_{FE}(E_{FE}) \cdot A = Q_{FE} = Q_{IF} = \varepsilon_0 \cdot k_{IF} \cdot \frac{A}{d_{IF}} \cdot (V - V_{FE}) \qquad (2.20)$$

follows

$$E = \frac{P_{FE}(E_{FE})}{\varepsilon_0 \cdot k_{IF}} \cdot \frac{d_{IF}}{d} + E_{FE} \cdot \frac{d_{FE}}{d}. \qquad (2.21)$$

2.4 Assessment of Non-ideal Ferroelectric Hystereses

Only for $t_{IF} \to 0$ (no interface layer) or $k_{IF} \to \infty$ (the interface layer equals a perfect electrode), the resulting curve of $P(E)$ exactly matches the curve $P_{FE}(E_{FE})$ (dashed curves in Fig. 2.6 c)). Lower k_{IF} cause broader and broader switching peaks mainly influenced by the second term $E_{FE} \cdot d_{FE}/d$ in Eq. (2.21). The first term containing $P_{FE}(E_{FE})$ accounts for the partial movement of the switching peaks beyond the P axis into the second and forth quadrant. In order to clarify what Eq. (2.21) implies, three special cases are considered:

1) When the polarization of the FE equals the remanent polarization, which represents the case of zero field across the FE ($P_{FE} = P_{FE}(E_{FE} = 0) = P_r$), the field across the complete stack is not zero but exhibits still a positive value. This means, in case of $E = 0$, E_{FE} will already be below zero. The ferroelectric is partially in a backswitching state without any field applied to the stack. The remaining polarization is thus, smaller than the P_r of the FE layer alone.

2) When the field across the FE equals the coercive field, its polarization becomes zero ($P_{FE}(E_{FE} = E_c) = 0$), the left term in Eq. (2.21) becomes zero and the right term shows, that $\|E\| < \|E_c\|$. The switching peaks are shifted toward lower fields and the hysteresis is narrowed by the presence of the dead layer.

3) As the absolute value of E_{FE} reaches its maximum P_{FE} also becomes maximized and therefore, both terms in Eq. (2.21) reach their maximum absolute values. The maxima and minima of E and E_{FE} are correlated.

A more detailed analysis was provided by Tagantsev et al.[56], who also outlined how characteristics of a real passive layer can be calculated from additional experimental data. Nonetheless, the simple approach presented here allows the conclusion: If a narrowed, rounded and slanted hysteresis with a reduced remanent polarization is observed, this very likely stems from the presence of a passive layer in series to the FE—typically at the electrode interfaces.

The above explanations are in agreement with Ma's derivations[57] for a metal-ferroelectric-insulator-semiconductor gate stack in a ferroelectric field effect transistor (FeFET). A "depolarization field" across the FE is caused by an insulating interface layer between the FE and the transistor channel and the major drawback for the data retention of the 1T cell concept of FE memories.

d) Fig. 2.6 d) top shows a reduced P_r as a result of **polarization fatigue**. Compared to case c), the curve is not rounded or slanted. The non-switching linear branches of the P-E curve remain unchanged. Also the position and width of the switching peaks is not changed. Only the peak area, i.e. the overall FE contribution, is reduced. Another difference to case c) to be aware of in practice: This curve shape evolves during an increasing amount of switching cycles as more and more domains stop taking part in the switching process. Polarization fatigue was discussed in more detail elsewhere[58, 59]. Common explanations include "domain wall pinning" (domain walls become immobile) and "seed inhibition" (the nucleation of nano-regions with inverted polarization is blocked).

e) The strongly constricted double-loop hysteresis shown in Fig. 2.6 e) top without any remanent polarization is typical for an **antiferroelectric (AFE)** material. At zero field, the opposed polarization of the two sublattices cancel each other. The unit cell is doubled compared to a FE phase and consists of two identical, but antiparallel oriented half cells.[60] As the absolute field increases, one of the halfs flips (peak I or II). The field only favors one of the polarization directions, as evident from Eq. (2.8). Decreasing the field again, the switched half cell flips back (peaks I' and II') into its original polarization state and cancels again the polarization of the second half cell.

The same macroscopic behavior can be generated by a field-induced phase transition from an originally unpolar unit cell into a polar unit cell due to the term $-P \cdot E$ in Eq. (2.8). This term always favors the polar state and can cause a phase transition if the energetic difference and the barrier between the non-polar and polar phases are low enough. Both, antiferroelectricity and **field-induced ferroelectricity (FFE)** equal the situation shown for the first-order phase transition in Fig. 2.5 ($T_C \leftarrow T$). The difference between both scenarios is a bit philosophic: In the AFE case the antiparallel FE unit-cell doubling exists, whereas in the FFE case, a completely different phase transforms into a FE phase. Nonetheless, the AFE phase is also a second, non-polar phase and the lattice constants of the half cells do not necessarily match those of the FE unit cell[61].

A third scenario can—at least from a modelling point of view—account for this hysteresis shape. As mentioned in section 2.3, different internal bias fields can exist throughout the film. If by chance, the exact same amount of FE material is biased towards positive and negative fields and $|E_{Bias}| > E_c$, such a double-loop hysteresis would be measured. Any intermediate cases would give rise to constricted hystereses with a non-zero P_r curves similar to what is shown in Fig. 2.6 f). In literature, hystereses ranging from case e) to f) were reported for unpoled, acceptor doped PZT[40], $(K_{0.99}Li_{0.01})(Nb_{0.65}Ta_{0.35})O_3$[62], $Na_{0.5}K_{0.5}NbO_3$[46], and $BiFeO_3$[63] as well as undoped $BiFeO_3$[47] samples. For the latter sample, elastic stresses induced by 180° domain walls led to the strongest backswitching (relaxation).[47] Unfortunately, no evolution of P-E and I-E with cyclic switching was shown. In contrast to the present case, the hystereses could be depinched by a thermal activation of 100 – 200 °C, which was sufficient to mobilize defects.[40, 46] Moreover, the maximum polarization increased, which is a hint on a reorientation of domains in the electric field or on the release of formerly stuck domains. The increase of P_r has to have a different origin, if no such increase in P_{max} is observed.

2.4 Assessment of Non-ideal Ferroelectric Hystereses

f) Similar to an AFE material as described in case e), a **ferrielectric** also exhibits a constricted hysteresis. But a certain P_r remains at zero field because the two sublattices do not cancel each other completely (Fig. 2.6 f) top).[64] A similar hysteresis can be achieved by a **mixture of FE and AFE** phase fractions.[64] For simplicity, it was assumed, that the fields to switch the FE and AFE domains to positive (FE ↑, AFE↑) and negative (FE ↓, AFE ↓) polarization interfere in the first and third quadrant. The field to switch back to zero net polarization (AFE 0) in the AFE unit cells are still found separate in the second and fourth quadrant. Coexistence of FE and AFE phases was reported for, e.g., lead barium zirconate[65, 66] and PZT[67]. Field-induced phase transitions were reported for $(1-x)Bi_{0.5}Na_{0.5}TiO_3 - xBaTiO_3$.[68] Tan et al.[69] even used terms like "relaxor ferroelectric", "weakly polar ferroelectric" and "relaxor antiferroelectric" to describe the different types of domains found in this material. Complex mixtures of these phases and transitions between them were reported. As a consequence, the sample itself was called "relaxor ferrielectric". Deepening the discussion about these terms is beyond the scope of this work and might become philosophic rather quickly since an equally short and comprehensive term is hard to find. However, it should be noted that such more complex situations could exist in the sample to be studied. Additionally, effects as described at the end of case e) could occur.

g) The important thing to be noticed in Fig. 2.6 g) bottom is the shift of the corresponding pair of switching and backswitching peaks towards the same side on the field axis. Carl and Hardtl[41] ascribed the occurrence of a constricted hysteresis and two distinct switching peaks per field polarity to different **internal bias fields** throughout the sample. In their formerly unpoled, aged PZT sample, the transient current peaks merged during field cycling. In Al doped PZT, a second phase around the grains was present. The aging process was attributed to space charges accumulating in this second phase and referred to as a "surface effect" or "grain boundary effect". No such phase was present for PZT doped with Mn or Fe. In these cases, the field cycling effect was speculated to result from two other effects or a combination of both—a so-called "domain wall effect" or a "volume effect". The latter involves electrical and mechanical dipoles (anisotropic centers) aligning in favor of a present spontaneous polarization in the respective domain. Thus, this polarization state is stabilized compared to the opposite state. Field cycling neutralizes the inhomogeneously distributed centers by space charges and the hysteresis "relaxes" (depinches, opens). The domain wall effect was attributed to the movement of dopant ions: Initially, the distribution is speculated to be homogeneous, which changes during aging. Electronic charges could be exchanged between different valence states or a diffusion to the domain walls could take place. A more equal distribution could be achieved by electric field cycling. However, domain wall and volume effect are hardly distinguishable by experiment. For either case, a restriction of domain wall mobility and a resulting decrease in dielectric losses is anticipated. Later studies, e.g. by Morozov and Damjanovic[38], also concluded that probably multiple phenomena are involved in practice.

h) Unlike in case g), Fig. 2.6 h) bottom shows corresponding peaks that posses the same distance to $E = 0$, i.e. represent domains with zero bias, but **different coercive fields**. This scenario war favored by Schenk et al.[70] to explain their observations in Sr:HfO$_2$. After having merged two switching peaks during a preconditioning step ("wake-up" effect[71, 42]), a split-up could again be induced by subjecting the sample to a cycling sequence at lower fields (non-saturated or subloop switching, referred to as "subcycling"). Even multiple peaks could be induced by multiple subcycling sequences from higher to lower amplitudes in any arbitrary configuration. The split-ups could be re-merged by field cycling at saturating amplitudes again. In agreement with Lines and Glass[31] and a mathematical modeling approach, the authors concluded that defects modify the local switching barriers of domains/regions and with it the respective E_c's. Field cycling redistributes the defects within the film. At saturating amplitudes, all domains switch and a uniform distribution results. At subcycling fields, the static (non-switching domains) offer a stable mechanical and electrical environment and accumulate defects, which additionally inhibit the switching of these domains in favor of the switching domains in which the amount of defects was reduced.

As tempting as this explanation was, later publications by the same group proved that a complex derivative of scenario g) was responsible for all observations. A detailed description of the respective experiments will be given in chapter 5 of this dissertation. Nonetheless, this does not rule out that a mechanism like this might dominate in another material.

A last scenario that could cause such plots are two dominating textures present in one sample. The easiest situation is half the sample consisting of domains with the polar axis perfectly aligned in direction of the applied field (i.e. normal to the electrodes) and half the sample exhibiting domains with the polar axis tilted, e.g. by 45° to the field. The lower projection of the field across the tilted grains result in a higher macroscopically measured coercive field. The merging of these peaks has been described as "pulse poling"[72]. In contrast to the conventional poling at elevated temperatures and applied bias field, it does not result in an imprint (see case b)), which is an advantage of this procedure. However, its manifestation differs from the case shown in Fig. 2.6 h):

- A "poling" in the sense of orienting P along the field lines, would only cause a peak movement from higher to lower fields. In the sketch, the lower-E_c peaks moves towards higher fields.
- The projection that cause the higher effective field for half the domains also results in lower effective polarization values (by the same factor). During poling, both maximum and remanent polarization should increase, which is also in contrast to what is shown in the sketch.

2.4 Assessment of Non-ideal Ferroelectric Hystereses

Together with the analog nature of the split-up/merging procedure, poling cannot account for the observations by Schenk et al.[70]. Similar to the explanation before, this does not rule out these effects for other samples, but the differences to what is sketched here, should be carefully considered when evaluating measurement results.

The field sensitivity of the FE phase mainly addressed in the scenarios e) to h) has also additional consequences. First of all, defects not only change the phase stability by representing a different element on a lattice place, but also by introducing additional charges and local electric fields. Secondly, polarization discontinuities occur at all interfaces of the ferroelectric. With the field generated by the uncompensated polarization charge, the ferroelectric destabilizes its own phase. Without sufficient shielding by electrodes, this leads to the characteristic domain pattern of regions with different orientation of the polarization, i.e. the unit cells of the ferroelectric phase. Phenomena related to the surfaces and uncompensated polarization charges were already recognized in the 1950s.[29, 73, 74] They were studied within the frame of phenomenologic theory of ferroelectricity as presented in section 2.2. Some of the derived effects include 1) a collapse of ferroelectricity at thicknesses below a certain critical value, 2) a conductivity increase of domain walls, and 3) the depolarization field dependence of the Curie temperature. For a collection of these phenomena, the reader is referred to the books by Lines and Glass (1977)[31] or Ghosez and Junquera(2006)[75]. The latter stated "[...] virtually nothing has been done concerning defects (such as oxygen vacancies, dislocations, or grain boundaries) and this constitutes certainly a challenge for the future." Thus, specific statements beyond what has been discussed in cases a) to h) above are difficult and depend on the present combination of material itself, its granular structure as well as the used electrodes, incorporated dopants and unintentional defects.

After this basic assessment, logical next steps could include the following more sophisticated studies[48]. Several of these methods have been used in the frame of this work as referenced below.

- First-order reversal curves (FORCs) [76, 77] to determine switching density plot (s. Preisach model in section 2.3) to distinguish different origins for the observed hysteresis shapes described above. Monitoring its change based on progressive poling, aging time, different process conditions, switching cycles, etc., can give further insights. → see section 3.6 (description) and sections 5.1 – 5.3 (results).
- TEM-based approaches such as HR-TEM (high resolution transmission electron microscopy) and STEM (scanning TEM) as well as X-ray diffraction[67, 65, 66, 68, 69] to identify phases compositions, stress fields and maybe their change with applied electric field. → see sections 3.2 and 3.4 (description) and different occasions of chapters 5 and 4 (results).
- Polarization microscopy[63, 78] (thicker samples) to study domain movement, identify 90°, 180°,... grain boundaries

- PFM (piezoresponse force microscopy)[79, 80, 81, 82] to study domain orientation, formation, kinetics, pinning, etc. → see section 3.3 (description) and section 5.5 (results).
- O_2 anneals (maybe with ^{18}O as radiotracer to monitor their incorporation); use of Ir, IrO_2, or other oxidic electrodes[22, 83, 84, 85]
- EPR (electron paramagnetic resonance)[86, 87], XAS (X-ray absorption spectroscopy)[88, 89], and Raman/FTIR (Fourier transform infrared) spectroscopy[90, 91] for different atomic coordinations, defect/dipole alignment, lattice distortions, phase transitions, etc.

2.5 Non-volatile Random Access Memory and Ferroelectric Memory

This section is devoted to introducing the concepts of ferroelectric memory and to deriving performance goals from a comparison to mainstream memory solutions.

First of all, memory hierarchy and its classic representatives of around 1980 should be considered[92]: The width of the hierarchy pyramid in Fig. 2.7 represents the data density. The cost per bit increases towards the top whereas the access time decreases. The central processing unit (CPU) directly addresses the random access memory (RAM). Therefore, the RAM has to be capable of being read and written at CPU speed, which is in the range of a few ns. Today, the fastest access times are realized by SRAMs in cache configuration. Data not directly used by the CPU is classically stored in magnetic disks with access times in the ms range. Long-term storage of data that is not used on a regular base is done on magnetic tapes as the cheapest option, but also with the lowest access times of several 10 s up to a minute. RAM represents the classical "volatile memory", whereas the magnetic disks or magnetic tapes are the classical "non-volatile" (NV) memory. Between these two types, an apparent "memory gap" of six orders of magnitude existed, that limited the performance of electronic devices. Two important key facts to remember:

- The term **"non-volatile"** refers to the capability of retaining data for **10 years**. (Note: This is a simplified statement sufficient for this work. In practice, temperatures and other boundary conditions have to be specified as well and differ for different applications.)
- A **"random access memory"** is a memory capable of **read and write at CPU speed**. (The term "random access", actually refers to the possibility of addressing any cell at any time to achieve the required speed, which necessitates the typical array arrangements of the cells in this type of memory.)

2.5 Non-volatile Random Access Memory and Ferroelectric Memory

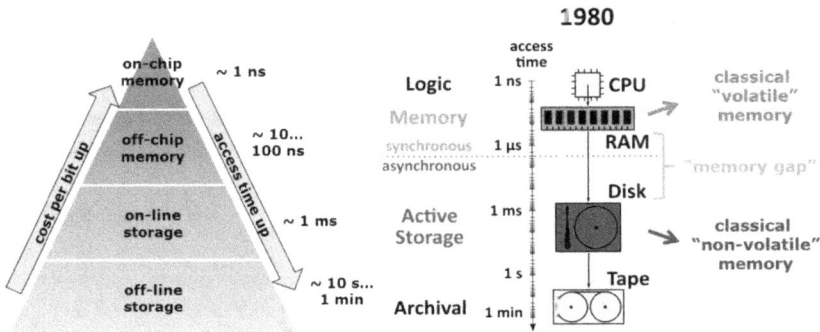

Fig. 2.7: Memory hierarchy (left) and classical representatives for the classes in the hierarchy (right). An apparent memory gap between classical "volatile" and "non-volatile" memory existed in 1980.[93]

In the past years, the situation changed (Fig. 2.8).[92] In 2009 the memory gap has been partially bridged by flash memories as they are used in USB (flash) drives and solid-state-disks (SSDs). In 2015, Intel and Micron announced the introduction of their so-called "3D XPoint Technology".[94] It is anticipated to become the first real "storage class memory" (SCM[2, 3, 95]), a class of memories supposed to completely bridge the memory gap. The term "storage class memory" was introduced in a 2007 press release by IBM as: "A new approach to creating faster storage, IBM's Storage Class Memory (SCM) research project is focused on creating low-cost, high-performance, high-reliability solid-state random-access storage that could compete with or replace disk drives and/or flash memory "[96] The question for the future is: Will there be a true non-volatile random access memory (NV-RAM) capable of replacing both RAM and hard disks in the initial scenario of around 1980 or will there be multiple specialized solutions bridging part of that span?

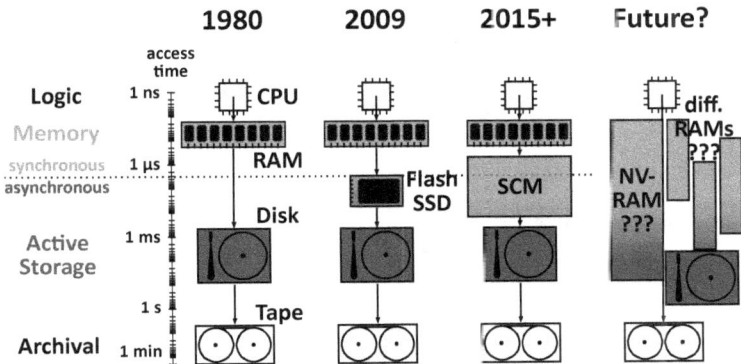

Fig. 2.8: Evolution of the main stream representatives of the memories classes.[93]

Challenges for NV-RAM arise from the so-called **voltage-time dilemma**[97]. The starting point for the discussion is the criterion of non-volatility— data retention of 10 y. To achieve this, a barrier of certain height is needed between the two memory states. However, in the switching case, this barrier has to be overcome in fewer than 10 ns. This means, we need a barrier that can be modified in a way guaranteeing both of these extreme cases, which are separated by more than 16 orders of magnitude on the time scale—10 y $\approx 3 \cdot 10^8$ s $\longleftrightarrow$ 10^{-8} s = 10 ns. A switching mechanism that scales in a highly non-linear manner with voltage is needed. However, compromise between high-speed and low-voltage is necessary, if the retention criterion is to remain untouched. In practice, the voltage-time dilemma can be extended by energy efficiency and the degradation of the barrier as an issue for unlimited cycling endurance. This interplay in the frame of an extended voltage-time dilemma (see Fig. 2.9) can be summarized as[93]:

1) For a non-volatile **data retention**, the barrier between the two states should be as high as possible. E.g. 1.5 eV barrier height was calculated to guarantee 10 y of retention at 85 °C for the case of an electron trapped between rectangular barriers of ≈5 nm base width.[98]
2) This increases the voltage necessary for sufficiently fast switching. Or the other way around, a higher barrier reduces the **switching speed** attainable by a given voltage.
3) A high field across the film stack accelerates the degradation due to defect movement/degradation finally resulting in earlier dielectric breakdown. For a ferroelectric, also fatigue mechanisms might be enhanced. The most extreme case of a 10 y life time with continuous switching operations in the 10 ns range defines the criterion for "unlimited" **endurance** as 10^{16} cycles.
4) At first glance, the argument for **energy efficiency** appears simple: For a given charge carrier, a higher voltage means an increased energy consumption $W = Q \cdot V$). However, the energy for the switching process itself is usually not the critical aspect. Charging the respective parasitic capacitances of bit-lines, word-lines, etc. in large arrays is a crucial aspects for energy efficiency. Another critical issue is the provision of high voltages by e.g. charge pumps in the peripheral circuitry.

The manifestation of this dilemma is differently severe for different types of memories (flash memory, phase change, resistive, magnetic RAM,...), but the general dilemma stays.

Before discussing the cell concepts of ferroelectric memory, a comparison to the performance of what is currently on the market is given. Tab. 2.1 shows performance characteristics for dynamic RAM (DRAM), flash memory and ferroelectric RAM (FRAM, FeRAM or FERAM). DRAM offers unlimited endurance and ns access time. The retentions, however, is only in the ms range. Flash, the second mainstream memory to be mentioned here, can guarantee a 10 y retention, but endurance and access time is compromised. FRAM combines the strengths of both devices offering non-volatility, unlimited endurance and ns access times.

2.5 Non-volatile Random Access Memory and Ferroelectric Memory

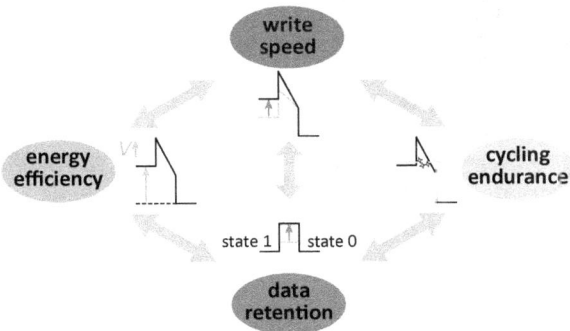

Fig. 2.9: Illustration of the extended voltage-time dilemma.[93]

Tab. 2.1: Comparison of mainstream memories and FRAM.

Memory	Endurance	Access Time	Data Retention	Technology Node
DRAM	unlimited	ns	ms	20 nm
Flash	$10^{4...5}$	ms (write)	10 y	16 nm
FRAM	unlimited	ns	10 y	90 nm

This means, NV-RAM already exists and the so-called FRAM is one example. However, the big drawback of FRAM is scaling. With the current 90 nm technology[2], it lags more than 10 years behind what is state-of-the-art for DRAM and flash.[17] Therefore, FRAM is no mainstream memory. After introducing the cell concepts, the reasons for these scaling issues and other major drawbacks will be derived. Moreover, it will be shown how HfO_2/ZrO_2-based ferroelectric could potentially overcome them.

Two types of cell concepts of ferroelectric memory are distinguished. A starting point for the discussion of these concepts could be the DRAM cell with its access transistor and storage capacitor as shown in Fig. 2.10. To make this cell an non-volatile memory cell, the dielectric in the capacitor can be replaced by a ferroelectric. This cell is referred to as 1T1C (one transistor one capacitor) cell. To spare the capacitor and reduce the cell size, the ferroelectric can be put directly on top of the transistor channel. Its polarization charge additionally depletes or populates the channel and turns the transistor off or on depending on the polarization state of the FE. This cell is called the 1T cell because it only consists of one transistor—a ferroelectric field effect transistor (FeFET). For a more detailed discussion, the reader is referred to refs. [22, 23, 101].

---
[2]FRAM was also implemented in 65 nm by Ramtron/Texas Instruments[99, 100], but no press releases or other publications stating a market introduction could be found.

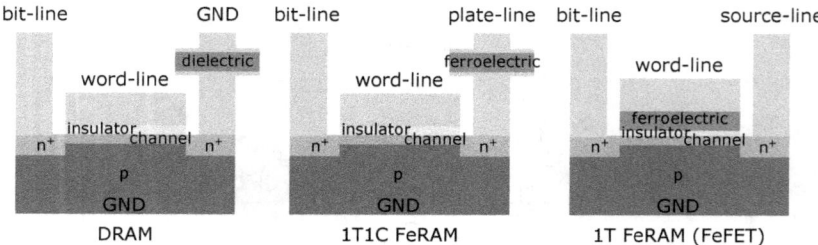

Fig. 2.10: Comparison of DRAM, 1T1C FeRAM and 1T FeRAM cells.

A basic understanding of the respective cell operation is inevitable to identify figures of merit and distinguish non-ideal, but also non-critical behavior from true roadblocks that certainly require engineering efforts.

The **1T1C cell** exploits the polarization charge stored in a capacitor as it has already been proposed by Buck in 1952[102]. His simple cross-point array of capacitors, however, suffered severe disturb problems, i.e. the states of the cells were easily destroyed by the voltage drops that occurred while addressing neighboring cells. With an additional access transistor, the issue is solved and the 1T1C cell is the cell concept used in commercial FRAM. To read the memory state of the capacitor via the bit-line (BL), the select transistor is opened via the word-line (WL). A defined pulse of always one and the same polarity is applied at the plate-line (PL). Depending on the switching state of the FE, the capacitor is either switched or remains in the same state. The current that is sensed as a potential change of the bit-line compared to a reference is either nearly zero (non-switching case: displacement current) or enhanced by the the switched polarization charge $2P_r \cdot A$ (switching case: displacement current and switching current). To enhance the reliability of commercial FRAM, a second 1T1C pair can be used. It always contains the opposite memory state and is used to generate the reference voltage for the sense amplifier. Such a 2T2C cell maximizes the sensed voltage difference and ensures a more stable read-out. Both half-cells suffer the same retention and cycling degradation, which causes an optimized read out through the whole life time of the device.

Scaling issues for the 1T1C cell:

- In the past, FRAM was limited to the 90 nm technology (see above) and the use of planar capacitors.[103] (ALD needed for 3D structures; complicated for a material with many metal components, that require a stable stoichiometry and the same ALD temperature window: PZT → metal precursors for Pb, Ti and Zr are needed)
- Dead layer effects hamper thickness reduction for conventional materials such as PZT.[55]
- Incorporation of heavy metals and "non-standard" chemical elements into CMOS fabrication are undesired due to processing and disposal issues.[104, 105]
  → 2010 projection of scaling: 65 nm technology node in 2022[104]

2.5 Non-volatile Random Access Memory and Ferroelectric Memory

The **1T cell** is switched (write operation) by applying a respective voltage between bulk and the WL at the gate. The read operation utilizes a fixed gate voltage at the WL in the middle of the memory window as defined by the threshold voltages of "0" and "1" state. Either a high ("on") or low drain current ("off") is sensed at the BL. The voltage difference between "0" and "1" is approximately given by $E_c \cdot d_{FE}$. It has to be high enough to ensure a clear current difference to properly define "on" and "off" for the sensing circuitry.

In addition to the material related scaling issues of the 1T1C cell, further restrictions arise from the necessity of an interface layer between FE and the channel, which usually exhibits a lower k-value than the FE[104, 105, 106, 57, 101, 23, 107]:

- The charge conservation in the capacitive voltage divider arrangement results in an extremely high field across the lower-k material, which degrades the interface layer. The fields in a metal-FE-insulator-semiconductor stack scale inversely with the k-values (see Eqs. (2.22) – (2.24)).
- Moreover, charge conservation gives rise to the occurrence of a depolarization field E_{depol}. The remanent polarization of the FE polarizes the interfacial dielectric. P_r causes a field of same sign across the interface layer. With zero net voltage applied to the stack, this results in a negative field drop across the FE, which counteracts P_r (see Eqs. (2.25) – (2.27)).
- The memory window directly scales with thickness. With coercive fields of $\sim 100\,\text{kV/cm}$, the target value of $\approx 0.5\,\text{V}$ (or better $1\,\text{V}$)[108, 109, 110] limits the thickness reduction of the FE layer to ~ 50 nm.

The problem of **increased field drop** across the lower-k interface layer can be derived from a simplified approach, which takes only the linear contributions into account (not P_r as done in Eq.(2.26)) and thus, represents only a lower limit:

$$Q = Q_{FE} = Q_{IF} = C_{FE} \cdot V_{FE} = C_{IF} \cdot V_{IF}. \tag{2.22}$$

Using the formulas for parallel-plate capacitor for the capacitances of the interfacial layer C_{IF} and the ferroelectric C_{FE}, one finds

$$\varepsilon_0 \cdot k_{FE} \cdot A \cdot E_{FE} = \varepsilon_0 \cdot k_{IF} \cdot A \cdot E_{IF}, \tag{2.23}$$

and canceling the capacitor areas A, which are assumed to be the same, the inverse proportionality of relative permittivities and electric fields is obtained:

$$k_{FE} \cdot E_{FE} = k_{IF} \cdot E_{IF}. \tag{2.24}$$

The **depolarization field** due to the dielectric interface layer is also derived from charge conservation:

$$Q = Q_{FE} = Q_{IF}. \tag{2.25}$$

This time, the remanent polarization of the ferroelectric film is included:

$$C_{FE} \cdot V_{FE} + P_r \cdot A = C_{IF} \cdot (V - V_{FE}). \tag{2.26}$$

For the retention case of zero external voltage V, rearranging yields the depolarization field

$$E_{depol} = \frac{V_{FE}}{d_{FE}} = -\frac{P_r}{\varepsilon_0 \cdot k_{FE} \cdot \left(1 + \frac{k_{IF} \cdot d_{FE}}{k_{FE} \cdot d_{IF}}\right)}. \tag{2.27}$$

2.6 Hafnium Oxide for Ferroelectric Memories

Hafnium(IV) oxide or hafnium dioxide with the chemical formula HfO_2 is the most stable and common oxygen compound of rare-earth metal hafnium[111]. Here and in general, it is mostly referred to as hafnium oxide or hafnia for reasons of convenience. Hafnia is chemically very inert[112], exhibits a high hardness[113], a density of about $9.68\,\text{g/cm}^3$[112] and a refractive index of 1.8 to 2.2[114] with low absorption in the visible spectrum, which make it interesting both for optical and protective coatings[114, 115].

HfO_2 and the chemically very similar ZrO_2 have been studied intensively as solid electrolytes (oxygen conductors)[116, 117] and as ceramics (with transformation toughening → "ceramic steel")[118] since at least the 1940s[119, 120]. At atmospheric pressure and room temperature, bulk HfO_2 exhibits a monoclinic phase of space group $P2_1/c$. Upon heating, it undergoes a phase transition into a higher symmetry tetragonal phase of space group $P4_2/nmc$ at about 1700 °C. A further increase of temperature finally stabilizes a cubic Fm3m phase at around 2600 °C before the melting point of 2800 °C is reached.[121, 122] Fig. 2.11 illustrates the respective unit cells of these phases. Further orthorhombic phases of space groups Pbca[123] and Pbcm[123, 124] have been reported to exist under various pressure and temperature conditions. Due to their oxygen conductivity, zirconia and hafnia have attracted interest in the fields of fuel cells[125, 126] and resistive switching[127].

In the semiconductor field, both oxides were investigated in detail as so-called "high-k" dielectrics to replace SiO_2. SiO_2 as the original gate dielectric no longer allowed a further thickness reduction as required by the scaling laws for the 90 nm technology node in 2003. With physical thickness reaching only 1.2 nm (= 5 atom layers), tunneling currents resulted in tremendous heat generation and power consumption.[19] To relax the scaling of the physical thickness for a given gate capacitance, materials with higher relative permittivity k ($k_{SiO_2} = 3.9$) became necessary. The requirements for such a material were summarized as follows[128]:

2.6 Hafnium Oxide for Ferroelectric Memories

- high enough k-value to be used for a reasonable number scaling nodes
- thermodynamically stable with Si
- compatible with thermal budgets of 1000 °C and 5 s
- good electrical interface with Si
- band offset to Si of $> 1\,\mathrm{eV}$ to minimize charge injection
- high band gap and low number of electrically active bulk defects to minimize leakage and trapping.

Here, problems automatically arise from the empirical relation of k-value to band bap ($E_{gap} \propto k^{-0.65}$) and breakdown field ($E_{BD} \propto k^{-0.5}$). Overall, HfO$_2$ provided the best compromise for transistors. The chemically similar ZrO$_2$ became state-of-the-art material for DRAM capacitors, where the temperatures to withstand are only around 550 °C.[97, 129, 130, 131]

In 2007, Intel presented their "High-k Solution" utilizing amorphous HfO$_2$ as gate dielectric and TiN as gate electrode material for the 45 nm technology node[19]. In the following years, there were efforts to further enhance the permittivity of hafnia by crystallizing it into a higher symmetry phase. Instead of k-values around 20 as for amorphous or monoclinic HfO$_2$[61, 132, 133, 134, 135], values of 28 to 70 for the tetragonal[61, 133, 135] and 29 to 39 for the cubic[61, 132, 133, 134, 135] phase could be possible.

In 2006, Toriomi et al. [136] reported a maximum in k for around 5 mol% of SiO$_2$ incorporation into HfO$_2$. Trying to make use of this maximum, Böscke and co-workers[1] found surprising humps in the small-signal C-E curves (see section 3.2 for a description of the method) and also a hysteretic behavior. This is a characteristic of a ferroelectric materials and thus, they also performed P-E measurements and checked the piezoelectric response via laser interferrometry. All these results pointed towards ferroelectric properties. The nonpolar Pca2$_1$ space group (at that time called Pbc2$_1$, which only is a different setting of the same group), as reported earlier for ZrO$_2$[123, 137] was proposed to explain this behavior.

In the following years, Y[132], Al[138], Gd[139], La[140], and Sr[141] doping as well as a solid-solution with the "sister oxide" ZrO$_2$[142] were successfully used to induce ferroelectric properties. Already in the first paper, the application in ferroelectric memories was pointed out.[1] Just one year later, in 2012, a ferroelectric field effect transistor in 28 nm CMOS technology was presented[143] supporting this initial claim. In 2014, the 3D capability was demonstrated by depositing Al:HfO$_2$ into trench capacitors[144], which resulted only in a negligible penalty in P_r compared to 100 % perfect area usage. The film properties were concluded to be comparable to those of planar films and also a comparable endurance of up to 10^{10} cycles was demonstrated at field amplitudes of 2 MV/cm. Lately, also first operational memory arrays were demonstrated.[145] In a 90 bit NAND array, the two states of all cells were clearly separated. However, this separation (memory window) was only 0.2 V, which is below the desirable levels of at least 0.5 V (or better 1 V)[108, 109, 110] as a consequence of a rather broad distribution of both "0" and "1" state. Besides the efforts for non-volatile random access memories, also non-volatile operation of µm-sized lab-scale transistors has been demonstrated[146]. The structure later on has been improved by adding an additional charge trapping layer on top of the FE to increase the retention from 100 s to 1000 s at 85 °C.[147] Also a negative capacitance effect in these devices has been concluded to underly the steep subthreshold slope.[147]. J. Mueller et al. contrasted the performance of capacitor and transistor-based memory devices in volatile (sub-loop switching) and non-volatile (saturated switching) operation.[107] To achieve RAM-like endurance performance with non-volatile retention behavior, additional engineering efforts are necessary. As already explained in section 2.5, the voltage-time-dilemma makes the NV-RAM still an ambitious aim.

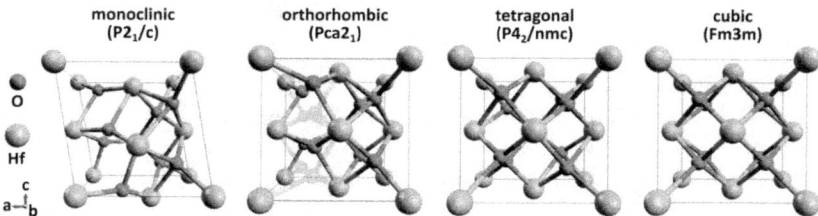

Fig. 2.11: Unit cells of monoclinic, orthorhombic, tetragonal and cubic hafnia polymorphs in (010)-view with lattice parameters taken from Materlik et al.[61]. (Avogadro version 1.1.1 and Avogadro² version 0.8.0[148] were used to the unit cell images.)

Despite the topic's novelty, already first PhD theses were at least partially devoted to it:

- Tim S. Böscke: "Crystalline Hafnia and Zirconia based Dielectrics for Memory Applications"[149]
- Johannes Müller: "[Ferroelectricity in Hafnium Oxide and its Application in Non-volatile Semiconductor Memories]"[150]
- Ekaterina Yurchuk: "Electrical Characterisation of Ferroelectric Field Effect Transistors based on Ferroelectric HfO_2 Thin Films"[151]
- Min Hyuk Park: "Novel material and device for ferroelectric memory: thin $Hf_{1-x}Zr_xO_2$ film and tri-states memory"[152]
- Stefan Mueller: "Development of HfO_2-Based Ferroelectric Memories for Future CMOS Technology Nodes"[153]

However, the industry-driven environment especially of the original group in Dresden (Germany) resulted in a strong device focus as reflected in the abovementioned titles. Additionally, typical films are challenging to study—polycrystalline, 10 nm thick and probably consisting of multiple phases. A basic notion of the origin of the ferroelectric behavior and of suitable levers to influence the ferroelectric properties existed. But solid proofs also in the context of recent theoretical work are still missing. Regarding the evolution of the ferroelectric hysteresis with cyclic switching, the status was similar: As important as this defect related behavior is, as challenging is a fundamental study. This thesis aims to fill these gaps and significantly advance the understanding of hafnia/zirconia based ferroelectrics.

3 Sample Preparation and Characterization Methods

3.1 Fabrication of Metal-Insulator-Metal Capacitors

Medium doped Si wafers (dopant concentrations of $10^{16}\,\text{cm}^{-3}$) were used as substrates. In the standard stack, TiN served as bottom electrode (BE) and was reactivly sputtered from a Ti target. On top of it, the actual ferroelectric layer was deposited by atomic layer deposition (ALD). The dopant concentration was varied by changing the amount of HfO_2 cycles between the respective dopant oxide cycles. A standard top electrode (TE) of TiN fabricated similar to the bottom electrodes completed the metal-insulator-metal (MIM) capacitor stack. Table 3.1 summarizes the process parameters for the respective stacks used in this work. After the deposition of the MIM stacks, the samples were annealed in N_2 atmosphere. Anneal times and temperatures are given in the respective sections of chapters 4 and 5. Samples with Si doping were fabricated solely at NaMLab by Claudia Richter, whereas TiN-Gd:HfO_2-TiN and TiN-Sr:HfO_2 stacks were provided by Christoph Adelmann and Mihaela Popovici, respectively, from Imec in Leuven, Belgium. For Gd and Sr doped HfO_2, the cationic ratio (dopant/[dopant + Hf]) was determined by elastic recoil detection analysis[154, 155] and Rutherford backscattering spectroscopy[154, 155], respectively at Imec in Leuven, Belgium. For the Si doped films deposited at NaMLab in Dresden, Germany, X-ray photoelectron spectroscopy (XPS)[154] was performed by Marion Geidel at IHM, Dresden.

On top of the MIM stacks, 10 nm Ti and 25 nm or 50 nm Pt were deposited via electron beam evaporation through a shadow mask. The resulting circular contact pads (diameters: 110 µm, 200 µm, 280 µm, 400 µm) for the prober needles served as hard mask to pattern the top electrode in a subsequent wet etch. A solution of H_2O, H_2O_2 and NH_4OH with a ratio of 50:2:1 (a diluted type of a standard clean 1, SC1 solution) at 50 °C was used for $\approx$ 5 min to selectrively remove the exposed TiN outside the Pt dots.

Tab. 3.1: Process parameters used to fabricate hafnia based MIM capacitors with different dopant materials. (RT = room temperature)

Dopant	Si	Gd	Sr
Bottom Electrode	TiN reactive sputtering at 250 °C	TiN or TaN reactive sputtering at RT	TiN reactive sputtering at RT
Ferroelectric	HfO_2 and SiO_2 ALD with $Hf(N(CH_3)(C_2H_5))_4/H_2O$ and $SiH_2(N(C_2H_5)_2)_2/O_2$-plasma at 280 °C	HfO_2 and Gd_2O_3 ALD with $HfCl_4/H_2O$ and $Gd(^iPrCp)_3/H_2O$ at 300 °C	HfO_2 and SrO ALD with $HfCl_4/H_2O$ and $Sr(^tBu_3Cp)_2/H_2O$ at 300 °C
Top Electrode	TiN reactive sputtering at 250 °C	TiN or TaN reactive sputtering at RT	TiN reactive sputtering at 200 °C

To establish an electrical connection, a high voltage was applied to short circuit two capacitors. A sufficient conductivity of this short was always checked via a leakage current measurement. One of these shorted capacitors served as the connection to the bottom electrode for one of the prober needles. The other needle was put on top of the working device to be characterized. This is the usual way that was used for electrical standard measurements. For piezoresponse force microscopy (PFM, see section 3.3) and impedance spectroscopy (IS, see section 3.5), a cleaved Si piece was glued to a Cu coated circuit board by conductive silver paint, which was also applied to the edges of the Si piece with the exposed bottom electrode. Fig. 3.1 shows a sketch of the resulting stack and the short to the bottom electrode.

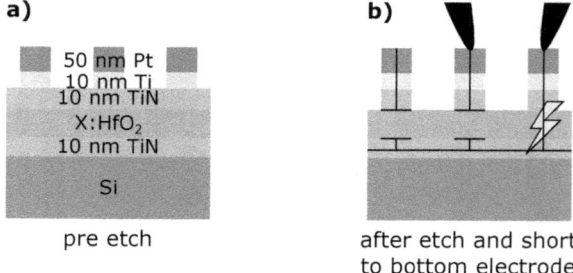

Fig. 3.1: Metal-insulator-metal stack: a) before and b) after SC1 etch with established short to the bottom electrode.(X: Si, Gd, Sr)

3.2 Basic Methods

The **basic structural characterization** included XRR and GIXRD. All measurements were performed on a Bruker D8 Discover with Cu-K$_\alpha$ ($\lambda = 0.154$ nm) radiation.

X-ray reflectivity (XRR) is similar to the approach of reflectometry with visible light.[156, 157] It allows determining (mass) density, thickness and roughness of the respective layers in a film stack. The main difference, is that instead of a broad wavelength spectrum and a fixed angle, only one characteristic wavelength of the X-ray tube is used an the incidence angle is varied. Fig. 3.2 shows a sketch of the expected intensity vs. 2θ (angle between tube and detector) plot expected for a very simple single layer. The position of the edge is defined by the (electron) density of the film, which determines its refractive index. Since the refractive index is always slightly below 1 in this wavelength range, total reflection occurs at the interface between air and the film for the lowest angles. It stays high until the critical angle given by the refractive indices is reached and X-rays start penetrating into the film. The period in the subsequent intensity modulation is caused by alternating constructive and destructive interference conditions. It is caused by the angle dependent difference in the optical path length of the two interfering beams reflected at the top and bottom surface of the layer. Thus, it mainly depends on the layer thickness. The intensity decay is defined by interface roughness and partial absorption of the X-rays in the layer, which increases with increasing incidence angles. This measurement is performed using a so-called Bragg-Brentano setup[158, 157]. It consists of a symmetric arrangement of the sample rotated to an angle θ and the detector rotated to 2θ with respect to the X-ray tube. Sample, tube and detector are placed at one and the same so-called focusing circle to yield a maximized intensity.

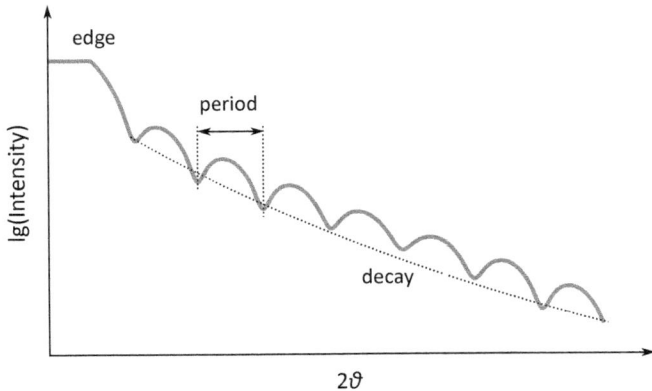

Fig. 3.2: Sketch of the features in a XRR plot for a single layer.[159]

X-ray diffraction (XRD) allows phase identification via characteristic peaks in a plot of intensity vs. 2θ, which are cause by constructive interference of beams diffracted at the present periodic arrangement of atoms in a crystal lattice.[157, 158] Instead of a symmetric

Bragg-Brentano setup, a grazing incidence geometry is used to limit the penetration depth of the beam and enhance the amount of information stemming from the first few 100 nm of the sample—i.e. the thin MIM stacks on top of the Si substrates in this case. As can be seen from a comparison of **grazing incidence X-ray diffraction (GIXRD)** and the classic Bragg-Brentano arrangement (Fig. 3.3), the lattice planes contributing to the measured signal are not parallel to the sample surface and their orientation continuously changes during the angular scan. This has to be considered when deriving statements about the texture or stress inside the sample. A list of the used reference patterns from the ICDD database is shown in Table 3.2

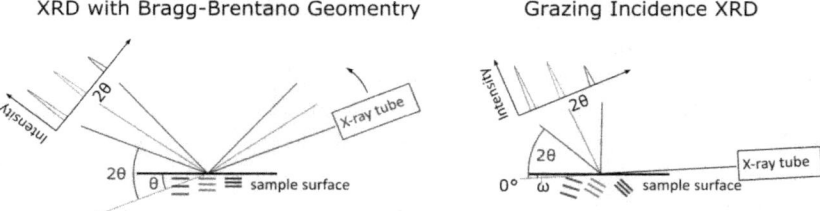

Fig. 3.3: Comparison of XRD geometries: a) Bragg-Brentano geometry vs. b) grazing incidence geometry.[159]

Tab. 3.2: Numbers of the PDF reference data (International Centre for Diffraction Data, PDF-4+ 2014, database version 4.1403) used for phase analysis.

Crystal Phase	Space Group	Dielectric Behavior	PDF Number
cubic HfO_2	$Fm3m$	paraelectric	04-003-2612
tetragonal HfO_2	$P4_2/nmc$	paraelectric	04-011-8820
orthorhombic HfO_2	$Pbcm$	paraelectric	04-003-6960
orthorhombic HfO_2	$Pbca$	antiferroelectric	01-083-0808
orthorhombic HfO_2	$Pmn2_1$	ferroelectric	[a]
orthorhombic HfO_2	$Pca2_1$	ferroelectric	04-005-5597
monoclinic HfO_2	$P2_1/c$	paraelectric	00-034-0104
cubic TiN	$Fm3m$	metallic	00-038-1420
cubic TaN	$Fm3m$	metallic	00-049-1283
cubic Pt	$Fm3m$	metallic	00-004-0802

[a]This phase was proposed by Huan et al.[160] as a result of ab-initio simulations.

The **basic electrical characterization** methods include polarization hysteresis (P-E) measurement and a small-signal capacitance (C-E) measurement. Background information on these and also on basic electromechanical characteristics of ferroelectrics can be found in an overview paper by Damjanovic.[161]

3.2 Basic Methods

In the measurement of **polarization vs. electric field** (*P-E*), the polarization is measured as integrated current versus the electric field, which is usually swept as a triangular waveform. The so-called virtual ground method[22, 162] implemented in an TF Analyzer 3000 by aixACCT Systems was utilized. Current is measured indirectly via an operation amplifier and a feedback resistor. The negative voltage necessary to keep the node at the input of the operation amplifier at 0 V, is proportional to the current flow through the feedback resistor (Fig. 3.4 c)). This setup has two advantages compared to the Sawyer-Tower[163], a setup sensing the voltage drop across a reference capacitor, and the shunt method[162] shown in Fig. 3.4 a) and b), a setup sensing the voltage drop across a reference resistor: 1) The complete excitation voltage drops across the ferroelectric capacitor. 2) Parasitic cable capacitances are made ineffective.[22, 162]

a) Sawyer-Tower method

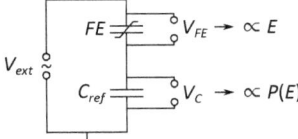

b) shunt method

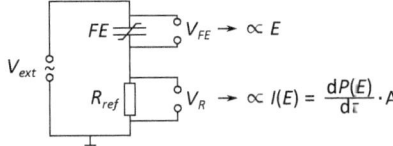

c) virtual ground method

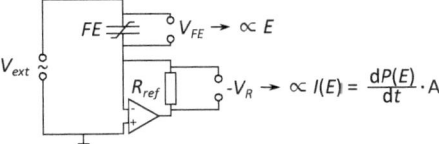

Fig. 3.4: Comparison of different setups for *P-E* measurements: a) Sawyer-Tower circuit, b) shunt method, c) virtual ground approach.[48]

The **small-signal capacitance as a function of applied electric bias field** (*C-E*) is measured via a small-signal sinusoidal continuous wave of 10 kHz frequency modulated on top of a bias voltage. Two methods were used in this work. In measurements with a 4200-SCS by Keithley Instruments, the bias field was stepwise increased with plateau widths of 2 s. In contrast, a triangular bias field sweep (large-signal) of 10 Hz frequency was used on the TF Analyzer 3000 by aixACCT Systems. The latter methods results in smoother *C-E* curves and a reduced stress for the sample, which allows using higher large signal fields.

3.3 Scanning Probe Microscopy

Scanning probe microscopy (SPM) utilizes a nano-scaled tip to probe surface properties via an atomic force microscope (AFM).[164] The lateral resolution is typically limited by the tip-radius usually in the range of a few 10 nm. In this work, piezoresponse force microscopy in single-frequency mode and in the form of band excitation point spectroscopy was used to measure local piezoelectric properties of the sample. Moreover, Kelvin-probe force microscopy in a just recently developed contact-mode variant was applied. A Cypher AFM by Asylum Research together with a custom-build data acquisition system (National Instruments: PXIe-6124, PXIe-1073, BNC-2110) and conductive ElectriMulti 75E tips by Budget Sensors were used for all SPM measurements.

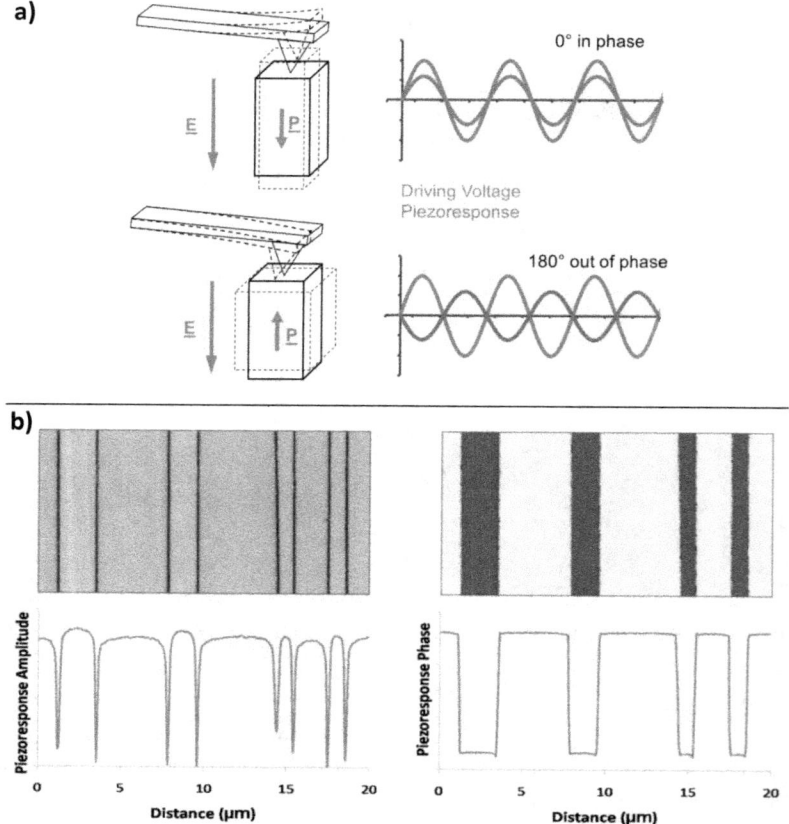

Fig. 3.5: a) Principle of SF-PFM with sinusoidal small-signal excitation and piezoelectric response. b) Typical amplitude and phase map with corresponding line profiles.[165]

3.3 Scanning Probe Microscopy

Piezoresponse force microscopy (PFM)[164, 79, 81, 82] makes use of the piezoelectric properties of a ferroelectric sample. As stated in section 2.1, every (crystalline) ferroelectric also exhibits piezoelectricity. Similar to the C-E measurement described in section 3.2, PFM represents a small-signal measurement. In **single-frequency PFM (SF-PFM)**, a small AC excitation voltage is applied between the bottom electrode and a conductive tip, which is in contact with the sample. This electric stimulus causes an expansion and contraction of the sample according to the converse piezoelectric effect as shown in Fig. 3.5. The height response is fed into a lock-in amplifier to only extract the component of the height response that follows the AC excitation with a stable phase angle to it. If the setup is calibrated, the phase and amplitude output of the lock-in amplifier represents the effective piezoelectric constant along field direction d_{eff} element in the piezoelectric tensor, i.e. the ratio of strain/electric field or dielectric displacement current/mechanical stress.[164] The amplitude and phase of the d_{eff} response depend on the applied external DC field as can be seen from Fig. 3.6. Sweeping the bias field, a local piezoelectric hysteresis can be recorded if domains are switched by this bias field. This approach is usually referred to as **switching spectroscopy** or point spectroscopy.[164] To enhance the signal to noise ratio, SF-PFM can be measured close to the resonance of the tip-surface contact. This resonance is usually in the range of some 100 kHz.

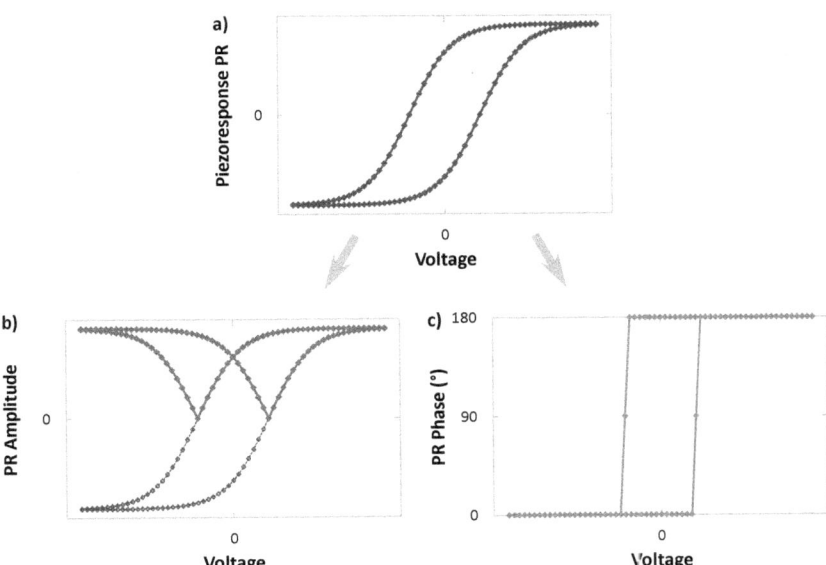

Fig. 3.6: Typical loops for switching spectroscopy: a) piezoelectric response (PR) for e ferroelectric sample consisting of b) amplitude (blue curve; unfolded violet curve equals PR shown in a)) and c) phase.

At zero field, the response depends on the polarization orientation of the material. The absolute value is identical, but the phase differs by 180° for domains pointing in opposite directions. This is utilized to image the domain patterns, as exemplarily shown in Fig. 3.5. At domain boundaries, the amplitude of the piezoelectric response vanishes and the phase flips by 180°. Differences in the amplitude of different locations on the sample surface can be due to different orientations (projections of the response and effective field) and locally different intrinsic and extrinsic properties. Also the tip-surface contact plays a role. Artifacts can occur due to topographies that either enhance or impede the coupling of the electric field to the sample. Moreover, if the leakage currents in the sample are high, they might compensate the electric field and no piezoelectric response is measurable.[166]

Very thin films, like the ones used in this work, rule out any measurements outside the frequency range of the contact resonance due to small absolute surface displacements. Unfortunately, the contact-resonance is not perfectly stable and might change over the scanned area by e.g. around 10 kHz or multiples of 10 kHz. This is, amongst other factors, due to changes in the contact area based on surface roughness or mechanical tip wear.[167] Since the excitation is fixed, this resonance shift results in a shift of the phase angle and amplitude. Since these two quantities are of interest to attribute piezoelectric/ferroelectric properties to the respective sample locations, such changes are undesired. In an extreme case, the excitation is left of the resonance at location one and right of it at location two. The two spots would give phases differing by 180°, which might be interpreted as opposite polarization states. Also the amplitude is strongly modulated when the excitation frequency moves relative to the contact resonance. A true piezoelectric response of the sample itself cannot be obtained reliably.

Band-excitation point spectroscopy (BEPS)[168, 167, 169] avoids these problems by evaluating a frequency band around the resonance. The frequency dependent piezoresponse is fitted by a simple harmonic oscillator (SHO) to extract the true sample related piezoelectric response. This could be done via the lock-in technique as before and sweeping the frequency at every single measurement point (of a typical 256×256 array), but it would result in extremely long measurements and undesired drift problems. To increase the speed of this approach, the desired frequency band around resonance is subjected to a Fourier transform. A corresponding pulse is generated by an arbitrary waveform generator and applied between tip and bottom electrode. An inverse Fourier transform is applied to the recorded cantilever response to obtain its reflection in the frequency regime to be subsequently fitted via the SHO model. Fig. 3.7 shows the schematic of this approach.

3.3 Scanning Probe Microscopy

a) Band Excitation and Response

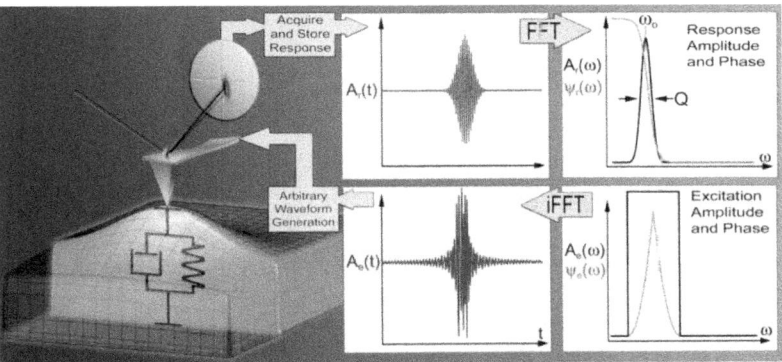

b) Simple Harmonic Oscillator (SHO) Model to Fit the Response

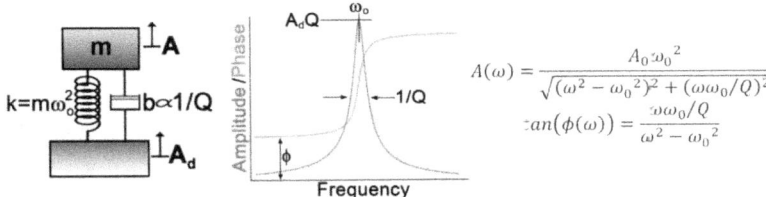

Fig. 3.7: Band excitation (BE) approach: a) Operational principle of the BE method in SPM. The excitation signal is digitally synthesized to have a predefined amplitude and phase in the given frequency window. The cantilever response is detected and Fourier transformed at each pixel in an image. The ratio of the fast Fourier transform (FFT) of response and excitation signals yields the cantilever response (transfer function). b) Schematic illustration of a SHO corresponding to mass on a spring with damping. Amplitude and phase versus frequency for the SHO model around resonance. SHO fitting yields amplitude, resonance frequency, and Q-factor, that are plotted to yield 2D images, or used as feedback signals. (adapted from [168, 167])

Recently, several publications addressed artifacts in PFM measurements that had not been in focus in previous years. **Contact-mode Kelvin-probe force microscopy (cKPFM)**[170] is one method that has been proven capable of supporting discussions on the influence of such spurious effects. Special attention is paid to the role of electrostatic forces between tip and sample and charge injection phenomena. Conventional Kelvin-probe force microscopy (KPFM, here described in amplitude modulation mode) is a non-contact technique. An AC voltage excitation V_{AC} is applied to stimulate an AC response of the capacitance between sample and tip. A first scan determines the height profile. In a second line scan, this height information is utilized to keep the distance between tip and surface constant.[170] The cantilever response consists of one component only dependent on the AC amplitude and another component dependent on the product of AC and the difference between bias voltage V_{DC} and the surface potential difference between sample and tip. Consequently, the surface potential difference can be extracted by sweeping the bias and detecting the minimum in the cantilever's AC response. For two ideal, chemically inert metals, the surface potential difference is given by the work function difference between tip metal and the metal at the sample surface. However, passivating layers can be present on the metals and not all surfaces of interest are metallic.

Locally different surface potentials also influence the outcome of PFM measurements. In case of PFM, not only the surface potential itself, but more precisely, the potential difference of the tip-surface junction (contact-mode!) is of interest. In most cases, PFM experiments are performed on the bare dielectric. Thus, a contact-mode equivalent of KPFM would be favorable. Moreover, a technique based on contact-mode would improve the lateral resolution, which in non-contact mode is limited by long-range Coulomb forces. Also in this work, measurements of capacitor structures, i.e. with the tip on a top electrode, were not feasible. The junction potential difference between the tip and a dielectric can be influenced by many factors, such as ionic motion, charge injection and electrochemistry. Therefore, electrical SPM methods might exhibit a strong time-dependence. Conventional KPFM is also not capable of monitoring these fast processes. Amongst others, this motivated the development of cKPFM, which will be described in the following.

A better time resolution is achieved by a stroboscopic approach of sampling the read voltage by different pulse levels for each of the write voltages used instead of slow bias sweeps as shown in Fig. 3.8. For each write voltage applied, a series of read voltages exists for which the AC response is recorded by a band-excitation approach as described before. From the response at the different read voltages, the junction potential difference present after the applied write pulse can be extracted as the x-intercept in the plot of cantilever amplitude vs. read voltage. For a ferroelectic material, the plots of cKPFM signal (like the piezoresponse) vs. read voltage is expected to show marked jumps around the coercive voltages (as will be shown in section 5.5). If this is not the case, the main contribution in PFM measurements unlikely stems from ferroelectric properties. The measured response is rather dominated by spurious artifacts as the ones described above.

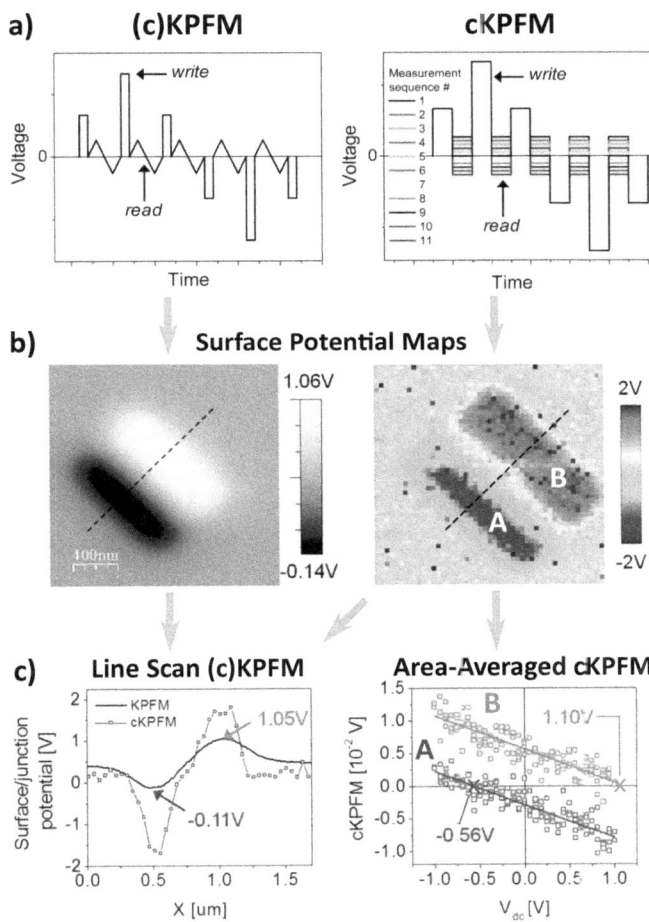

Fig. 3.8: Comparison of KPFM and cKPFM: a) Applied excitation signals for read and write. KPFM uses a bias sweep to determine the surface potential, whereas a stroboscopic approach (repetition of the bias pulsing sequence with different read voltage levels) is used for cKPFM. b) Surface potential maps from KPFM and cKPFM measurements after charge injection with ±5 V applied to the scanning tip. c) Line scans across the two regions via KPFM and cKPFM in the left graph and averaged cKPFM curves from the two differently poled areas in the right graph. The area-averaged cKPFM graph also illustrates how the surface potential is obtained from as the x-axis intercept, where the cKPFM signal becomes zero. (adapted from [170])

3.4 Transmission Electron Microscopy

The transmission electron microscopy (TEM) measurements for this dissertation were performed by:

- Xiahan Sang and Everett D. Grimley at North Carolina State University in Raleigh, NC using a probe-corrected FEI Titan G2 60–300 kV operated at 200 kV.[171]
  → RevSTEM, PACBED
- Darius Pohl at Leibniz IFW using a FEI Titan[3] 80–300 microscope operated at 300 kV (equipped with a field emission gun, CEOS CetCor CS-image corrector and a high-angle annular dark field detector).[172] → HR-TEM, STEM

The lateral resolution of microscopy is fundamentally limited to a value of about half the wavelength used for imaging (Abbe diffraction limit).[173] With visible light, this limit is in the range of 200 nm, which is insufficient to assess lattice structure of a crystalline sample. Atomic distances are in the range of a few Å ($0.1\,\text{nm} = 10^{-10}\,\text{nm}$), which necessitates the use of radiation with wavelengths also in the Å- or sub-Å-range. X-rays (e.g. Cu-K_α: 0.154 nm) or electrons (e.g. 200 kV: 2.51 pm) are examples of suitable types of radiation.

Transmission electron microscopy (TEM) is a well-established technique to study samples in microelectronics down to atomic resolution. A great variety of different TEM approaches exists in literature[174]. Most commonly, TEM refers to the following imaging approach: A coherent, parallel electron beam transmitted through the sample interacts with it and forms a complex coherent image. The contrast of that image is based on the crystal structure, the atomic mass of the structures and the thickness. Focusing the image or in the back-focal/diffraction plane on the camera, a magnified real space image or a diffraction pattern can be recorded. Both images represent the Fourier transform of the other.[174] An important numerical approach to be mentioned here, is the fast Fourier transform (FFT)[174]. In order to transmit electrons through a sample, a thin lamella needs to be cut. The desired sample thickness can vary based on the specific test, but it is generally desirable for lamellas to be below ≈50 nm. Else the intensity becomes too low due to absorption and inelastic scattering. In polycrystalline samples, grain overlap becomes more likely, which creates interference patterns (Moiré patterns[174]) that hamper structure analysis. Nonetheless, a certain thickness is needed to obtain a sharp diffraction pattern.

High-resolution TEM (HR-TEM) puts additional challenges to the microscope to achieve atomic resolution, such as beam-size and stability as well as correction of imaging errors due to lense imperfections (abberations), to make use of the phase contrast. This contrast is an interference pattern of the electrons on the camera and requires sophisticated methods of image reconstruction because the interference pattern is no direct image of the sample.[174] Both the real space and the diffraction image allow for identification of atomic distances and the reconstruction of the crystalline phase. Lower-index crystallographic axes ([100], [110], [111],... rather than [333], [435] or similar) exhibit wider atomic distances lowering

3.4 Transmission Electron Microscopy

the resolution requirements into a feasible range. In polycrystalline samples, the orientation usually differs from grain to grain and phase identification is limited to grains that can be oriented with a lower-index axis towards the beam within the tilting range allowed by the stage of the respective electron microscope. Apertures in the diffraction or the image plane allow selecting or excluding certain sample areas or diffraction maxima (families of lattice planes) before recording an image.

In contrast to conventional TEM, **scanning transmission electron microscopy (STEM)** utilizes a converged probe (focused, incoherent electron beam) that is moved across the sample pixel by pixel. Any significant electronic intensity measured at any location inside the sample, result from the very small sample area probed. This setup allows for a variety of detectors 1) in the direct beam (bright field, BF, 2) in a small angle (below 40 mrad, annular bright field, ABF)[175] or 3) in a higher angle to the direct beam (annular dark field, DF). "Annular" refers to the fact that ring-shaped detectors located around the direct beam are used for these image modes. Going to higher angles (inner angle above 50 mrad, **high-angle annular dark field, HAADF**)[176], can be advantageous because the scattering intensity of the electron is often approximately proportional to the square of the atomic number (Z) making the Z contrast the dominating contrast mechanism in the images.[174] Besides the Z contrast, thermal diffuse scattering is the second main contribution to image contrast.[177, 178, 179] STEM is a raster technique over nanometer sized sample areas. Thermal drift of the sample, sample holder and microscope can cause serious image distortions hampering the assessment of the crystallographic structure. In this work, the drift-corrected approach of **revolving STEM (RevSTEM)** (see Fig. 3.9) was used. Fast line scans rule out drift within the lines. Drifts between the lines are eliminated by scanning multiple frames at different rotation angles to accurately assess the drift vector and remove it from the image.[180] This method has been used to determine accurate lattice constants in section 4.2.

Annular detectors do not block the direct beam, which can therefore be utilized for **electron energy loss spectroscopy (EELS)**[174]. EELS uses element-specific absorption bands, which are a fingerprint of the electronic shells of the respective chemical elements in the sample. Moreover, as it is usually done in scanning electron microscopes, an X-ray detector can be used to do **energy-dispersive X-ray spectroscopy (EDX or EDS)**. EDX measures the energy and intensity of X-rays generated by relaxations of electrons between different energy levels, which have been excited by the electron beam. In SEM the signal originates from a bulb-shaped volume in the sample with a depth and diameter in the µm-range or even higher for higher-energy electrons. Since thin lamellas are used in STEM, this bulb-shaped interaction volume is cut before it starts widening and the problem of lateral resolution relaxes. Special instruments even allow for "atomic resolution EDX".[181] Similar to the use for the TEM imaging techniques, this atomic resolution capability refers to distinguishing elements atom-column-wise, i.e. laterally, and not within one atom column in direction of the depth of the lamella.

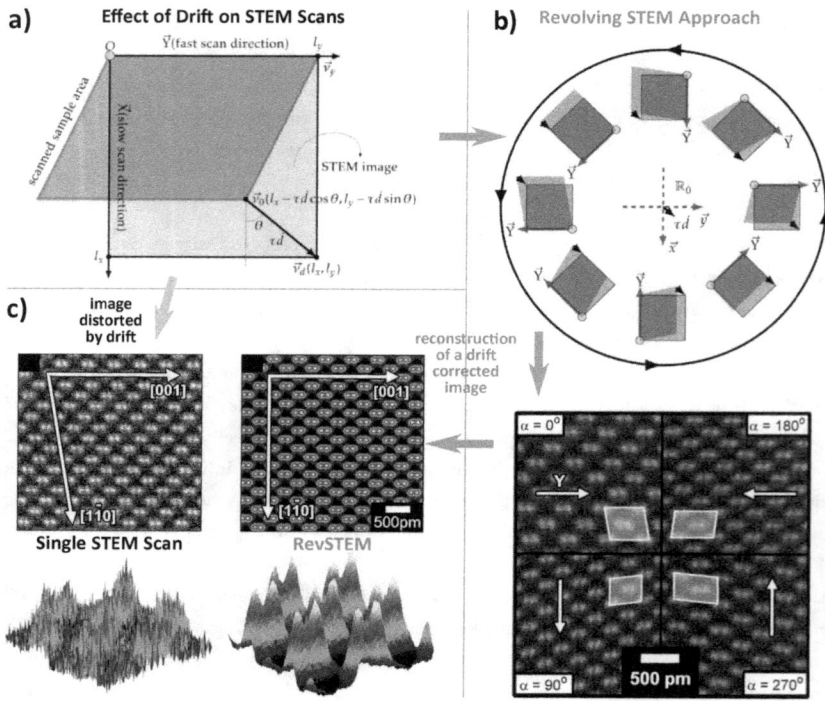

Fig. 3.9: Sketch of the RevSTEM approach: a) influence of drift in STEM images, b) measurement of different frames rotated around the image normal; c) uncorrected raw image from a single scan (left) and drift corrected image reconstruction (right). (Figure consists of different images by LeBeau et al.[180])

As explained at the beginning, conventional TEM is a diffraction-based approach. To assess more details of the crystalline unit cell, multiple diffraction directions can be utilized at once by a convergent instead of a parallel beam. Moreover, the converged beam provides much more localized information of the sample.[182] This technique is called **convergent-beam electron diffraction (CBED)**. The sharp reflexes given by the Bragg equation (similar to XRD described in section 3.2) widen to discs. The numerical assessment of the resulting intensity pattern of overlapping discs is more complicated than for parallel illumination, but includes potentially more information about the symmetry of the unit cell of the crystal. This method—in a position-averaged flavor called **PACBED**—has been used to evidence a missing centrosymmetry to experimentally prove the existence of a ferroelectric phase in section 4.2. The term "position-averaged" emphasizes that an averaged CBED pattern is recorded with a probe scanned across the sample. Tab. 3.3 contrasts the conventional electron diffraction with CBED and PACBED. As the semi-convergence angle of the electron beam becomes wider, the spots in the diffraction plane are enlarged to discs. For simplicity the intensity modulation within the discs is not shown. It is obvious that any asymmetry in the crystal is reflected

3.4 Transmission Electron Microscopy

in the pattern of overlapping discs and can be assess simply be visual inspection.[182] One main advantage of the PACBED compared to conventional CBED is that the same semi-convergence angle of the Å-sized probe as for the corresponding STEM images can be used without significant readjustment efforts. To still cover a certain desired area of the sample, the beam is scanned over the lamella. In practice, a rectangular area is chosen in a way that it is large enough to assess a complete unit cell but still small enough to not eliminate local changes by averaging. Besides polarity, the obtained diffraction pattern is highly sensitive to sample thickness, tilts and the semiconvergence angle (Fig. 3.10)

Tab. 3.3: Comparison of electron diffraction methods: conventional electron diffraction, convergent beam electron diffraction (CBED) and position averaged convergent beam electron diffractions (PACBED).

Method	Conventional Electron Diffraction	CBED	PACBED
Beam Shape	parallel	covergent (lower semi-convergence angle)	covergent (semi-convergence angle similar to STEM)
Sketch of Incident Beam on the Sample	beam → sample, probed grain on zone axis with corresponding lattice planes		scan of the desired area
Diffraction Plane	spots (200) (210) (220) (100) (110) (120) (000) (010) (020)	separate discs	overlapping discs

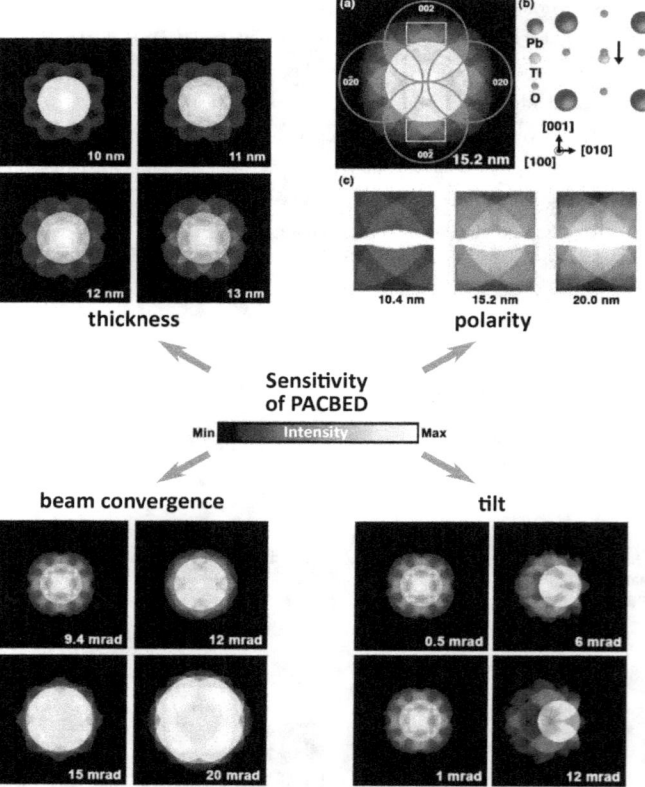

Fig. 3.10: Examples of simulated patterns to show the sensitivity of PACBED to sample polarity (PbTiO$_3$), thickness, tilt and the choice of the semi-convergence angle of the electron beam (all SrTiO$_3$). (Figure consists of different images from [182])

3.5 Impedance Spectroscopy

Impedance Spectroscopy (IS) is a well-established technique to identify structural inhomogeneities in samples via frequency dependent small-signal measurements of the complex impedance $Z = Z' - Z''$ (prime and double prime indicate real and imaginary part, respectively).[183] It is based on modeling the dielectric AC response of a system via an equivalent circuit. The properties of the system are extracted as the fit parameters of the selected equivalent circuit. First the measurement setup as well as the measurement approach are explained and later on, the mathematical descriptions of the circuit elements important in this work are given.

A 4294A Precision Impedance Analyzer by Agilent Technologies was used. A small-signal amplitude of 150 mV was used to have the same field as for the small-signal C-E measurements

3.5 Impedance Spectroscopy

for the 27 nm thick sample used for IS. A point averaging factor of 25 was applied and the analyzed frequency range was 40 Hz to 1 MHz. Coaxial Force-and-Sense cables of 2 m length are used. One side of these cables is connected directly to the low and high force and sense ports of the impedance analyzer. The other ends of the corresponding pair of force and sense cables are connected just in front of the needles within the manipulators of the probe station. A short cable connects the grounds of both manipulators directly at the measurement site. The excitation voltage is forced to the bottom electrode, which is insulated from the chuck of the probe station by an anti-slide mat made of rubber. The current response is sensed at the bottom electrode of the samples. Phase, short, open and load compensation are performed before the measurement as described in the manual of the 4294A.[184]

The approach of impedance analysis assumes a linear response of the system, i.e. the applied sinusoidal voltage excitation $V(t)$ results only in a sinusoidal current response $I(t)$ of same angular frequency ω and shifted by a certain phase angle φ (Eq. (3.1)). Fig. 3.11 summarizes the basics of **linear response theory** including the special cases of purely capacitive, resistive and inductive responses. In order to ensure a linear response, the excitation signal should be as low as possible. However, a reasonable signal-to-noise ratio, requires a sufficiently high voltage amplitude.

$$V(t) = V_0 \cdot e^{j\omega t} = V_0 \cdot \sin(\omega t) \quad \longrightarrow \quad I(t) = I_0 \cdot e^{j\omega t + \varphi} = I_0 \cdot \sin(\omega t + \varphi) \quad (3.1)$$

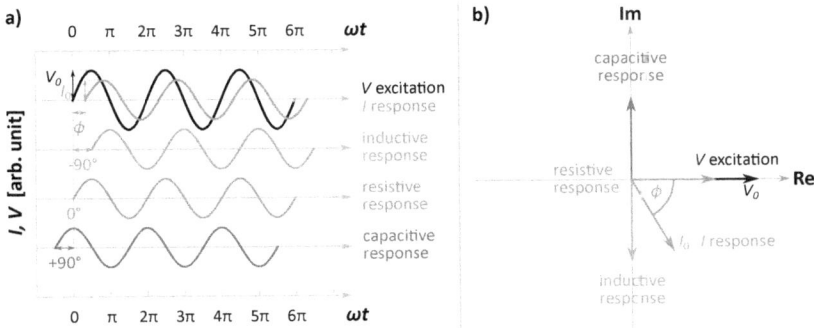

Fig. 3.11: Linear response theory for fixed voltage excitation (admittance perspective): a) sinusoidal excitation and response signals; b) corresponding phasor/pointer representation.[185]

A dielectric is supposed to act as a perfect insulator and is thus, only represented by a capacitance C_p. However, in real materials a certain finite conductivity always exists, which requires the introduction of a resistance R_p in parallel to the capacitance. Moreover, another resistance in series to this RC element is often found in practice. It stems from non-ideal electrodes, electrode interfaces or non-compensated (contact) resistances in the circuit. Usually R_s is at least one order of magnitude lower than R_p (ideal case: only C_p needed, because $R_p \to \infty$ and $R_s \to 0$). Fig. 3.12 b) shows this very basic equivalent circuit in the inset of the graph. In the following, the important plots and features in these plots are explained as they are fundamental to understand how circuit analysis via IS works and how Z evolved with the frequency.

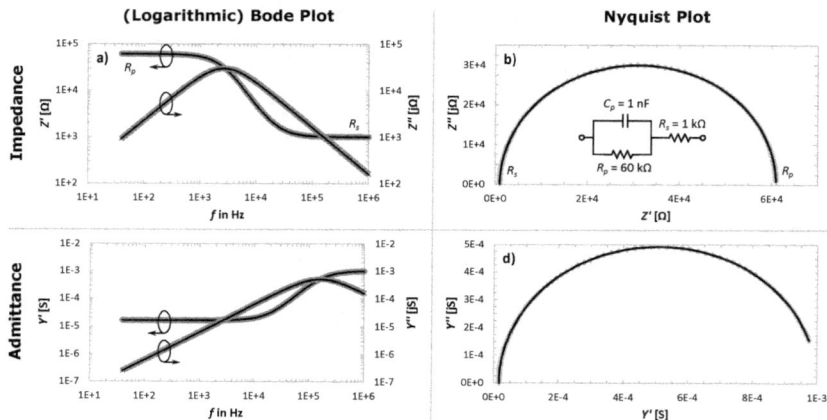

Fig. 3.12: a) – d) Simulated impedance spectroscopy results for the simple equivalent circuit (inset of b) discussed in the text: Bode (real and imaginary part vs. frequency) and Nyquist plots (locus, i.e. imaginary vs. real part) of impedance and admittance.[185]

Starting at lowest frequencies ($\omega \to 0$), the capacitance does not permit any current flow (see. Eq. (3.4)). Thus, all measured current has to pass R_p (independent of ω, see Eq. (3.2)), which is higher than R_s and thus, dominates the remaining circuit by limiting the allowed current flow. Increasing the frequency, current starts flowing also through C_p, which is out of phase to the current flow through R_p. An imaginary component of admittance Y and impedance Z arises. For even higher frequencies, C_p short-circuits R_p completely and allows for very high current flows. These high currents are only limited by R_s across which a higher and higher fraction of the excitation voltage drops. Thus, in the high frequency regime ($\omega \to \infty$), the circuit is dominated by R_s and the imaginary component of admittance/impedance decreases again. These features can be identified in Fig. 3.12 a) and c) in the so-called Bode plots (here, with logarithmic y-axis) of real and imaginary impedance/admittance components vs. (angular) frequency. Another type of characteristic plots for impedance analysis is shown in Fig. 3.12 b) and d): A plot of the imaginary vs. the real part of impedance/admittance is called a Nyquist plot. As a result of what has been explained for the respective Bode plots, these loci exhibit the shape of perfect semi-circles (referred to as "relaxation").[183]

In practice, the limited frequency range and the difference between the values of R_s and R_p results in incomplete circles of measurement data points. Moreover, more than one relaxation might be observed, which hints at a more complex composition of the sample. In

3.5 Impedance Spectroscopy

ceramics, often two circles are observed: One for the RC combination of the grain volume and a second RC combination for the grain boundaries.[183] These relaxations might also overlap depending on the parameters of the two (or more) RC elements. Last but not least, another deviation from ideal behavior is to be mentioned. Up to now, only combinations of ideal resistive and ideal capacitive properties were considered, which fully obey Eqs. (3.2) to Eqs. (3.5). Inhomogeneities, such as lateral or vertical gradients in both R and C or film roughness were found to result in an electrical behavior described by the so-called "constant-phase element" (CPE)[183] as described by Eq. (3.6) and Eq. (3.7). The formula for the CPE can be derived e.g. from an infinite parallel connection of R and C series arrangements to account for inhomogeneities mentioned above. An **overview of the circuit elements** used in this work is given below.

A linear **resistor** is characterized by

$$Z = Z' - Z'' = Z' = R \tag{3.2}$$

as well as

$$Y = Y' + Y'' = Y' = \frac{1}{R}. \tag{3.3}$$

A linear **capacitor** is described by

$$Z = Z' - Z'' = -Z'' = -j\frac{1}{\omega C} \tag{3.4}$$

and

$$Y = Y' + Y'' = Y'' = j\omega C. \tag{3.5}$$

The impedance and admittance of a **constant-phase element** are given by

$$Z = Z' - Z'' = Q = \frac{1}{T \cdot (j\omega)^n}$$
$$= \frac{1}{T \cdot \omega^n} \cdot \frac{1}{\cos\left(n \cdot \frac{\pi}{2}\right) + j \cdot \sin\left(n \cdot \frac{\pi}{2}\right)} = \frac{1}{T \cdot \omega^n} \cdot \left[\cos\left(n \cdot \frac{\pi}{2}\right) - j \cdot \sin\left(n \cdot \frac{\pi}{2}\right)\right] \tag{3.6}$$

and

$$Y = Y' + Y'' = T \cdot (j\omega)^n$$
$$= T \cdot \omega^n \cdot \left[\cos\left(n \cdot \frac{\pi}{2}\right) + j \cdot \sin\left(n \cdot \frac{\pi}{2}\right)\right], \tag{3.7}$$

respectively. Here, T is often referred to as the pseudo-capacitance because it equals C just for $n = 1$. The second parameter of the constant-phase element, n, is both exponent of the term $(j\omega)$ and a pre-factor representing the fractional deviation from the 90° phase angle of an ideal capacitance. A look at Eq. (3.8) reveals why this circuit element is called "constant-phase" element:

$$\frac{Z''}{Z'} = \frac{Y''}{Y'} = \frac{\sin\left(n \cdot \frac{\pi}{2}\right)}{\cos\left(n \cdot \frac{\pi}{2}\right)} = \tan\left(n \cdot \frac{\pi}{2}\right) = \text{const.} \qquad (3.8)$$

The phase angle is only determined by n and remains constant throughout the whole frequency range. Mathematically, it is allowed to be any real number. Three special cases are to be distinguished to show that the element is capable of exhibiting ideal 1) inductive (inductance L), 2) resistive or 3) capacitive behavior:

$$Z = \begin{cases} Z'' = j\omega L & \text{, for } n = -1 \rightarrow \text{ ideal inductor} \\ Z' = R & \text{, for } n = 0 \rightarrow \text{ ideal resistor} \\ Z'' = (j\omega C)^{-1} & \text{, for } n = 1 \rightarrow \text{ ideal capacitor.} \end{cases} \qquad (3.9)$$

3.6 First-Order Reversal Curves (FORC)

A recent review[48] on reasons for hysteresis deformation stressed the advantages of a closer look on the transient current curves (see section 2.4) as a first step toward finding the root causes for this deformation). First-order reversal curves (FORC) measurements were suggested as a next useful step to distinguish between different potential scenarios.

According to the Preisach-model explained in section 2.3, a ferroelectric can be understood as consisting of an ensemble of bistable units. Each of these units exhibits a certain field for switching and backswitching. The integral over all of these contributions gives the macroscopically measured polarization loop. A frequency plot of switching and backswitching fields is called switching density or Preisach density plot. FORC measurements were originally used to study ferromagnetics[186], but have also been adapted to conventional ferroelectrics in the past decades.[76, 77] The FORC approach makes use of the so-called "wipe-out" property of the Preisach model: In saturation, all history dependence of the dielectric properties of the ferroelectric is gone and only one and the same polarization state exists every time this saturating field is present. Thus, a saturating field E_{max}—in this work a positive one—is the starting point of the following series of FORCs. The electric field is swept to a lower field and back to saturation again. This lower voltage is called reversal field E_r and is subsequently decreased in equidistant field steps until the negative saturation field $-E_{max}$ is reached. During the ascending branches of the FORCs, transient currents and corresponding polarization curves $P_{FORC}^-(E_r, E)$ are recorded as exemplarily illustrated in Fig. 3.13. The switching density represents the mixed second derivative of the polarization response as shown in Eq. (3.10). For the sake of clarity, an easy numerical deviation of the switching density is explained first: The difference in the switched polarization during the ascending field sweep is the result of more and more domains being switched toward negative polarization

3.6 First-Order Reversal Curves (FORC)

by the stepwise decrease of the reversal field. Thus, the difference between two subsequent transient current density curves $\Delta I(E_{r,i}, E) = I(E_{r,i}, E) - I(E_{r,i-1}, E)$ is proportional to $\rho^-_{FORC}(E_{r,i}, E)$:

$$\rho^-_{FORC}(E_{r,i}, E) = \frac{1}{2} \cdot \frac{\partial^2 P^-_{FORC}(E_r, E)}{\partial E_r \partial E} = \frac{1}{2 \cdot \dot{E}} \cdot \frac{\partial j^-_{FORC}(E_r, E)}{\partial E_r}$$

$$\approx \frac{1}{2 \cdot \dot{E}} \cdot \frac{j^-_{FORC}(E_{r,i}, E) - j^-_{FORC}(E_{r,i-1}, E)}{E_{r,i} - E_{r,i-1}} \quad (3.10)$$

The unit of $\rho^-_{FORC}(E_{r,i}, E)$ is C/V^2. It can be plotted as z-value for each i-th reversal field $E_{r,i}$ on the y-axis versus the electric field E with $E_{r,i} \leq E \leq E_{max}$ on the x-axis. With increasing running index i, the area of the right plot in Fig. 3.13 is linewise filled below the diagonal $E_r = E$ from top ($E_{r,i} \approx E_{max}$) to bottom ($E_{r,i} = -E_{max}$). Each additional polarization contribution at a certain switching field E is a result of additional hysterons being backswitched at $E_{r,i}$. Thus, the coordinates (E_r, E) equal (E_{bs}, E_s) within the resolution of the E_r and E sampling.

For a better comparison to the common terms of coercive field E_c and bias field E_{bias}, the following coordinate transform can be applied:

$$E_c = \frac{E - E_r}{2}, \quad E_{bias} = \frac{E + E_r}{2}. \quad (3.11)$$

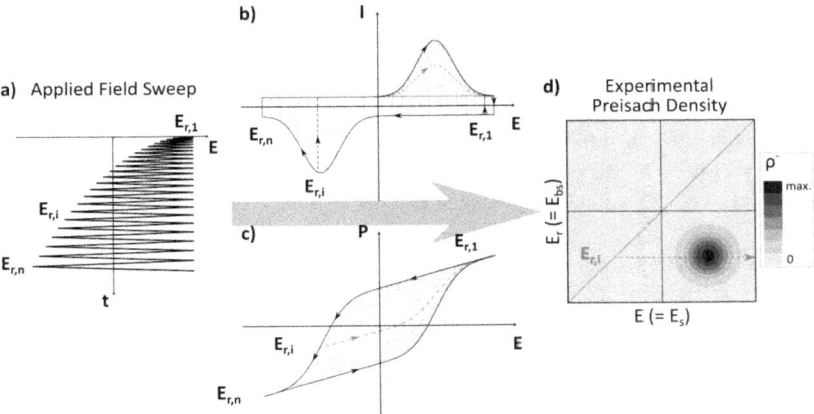

Fig. 3.13: Illustration of the FORC approach: a) excitation waveform, resulting b) I-E and c) P-E response and d) reconstructed switching density. ($E_{r,1}$ and $E_{r,n}$ are the first and last reversal field of the FORC measurement respectively.)[187]

The FORC measurements at NaMLab were established by Michael Hoffmann within his research project[188] as part of the master course Electrical Engineering at TU Dresden. A Keithley 4200-SCS equipped with pulse generator and oscilloscope was used. FORC sweeps were applied via the pulse generator and the current was measured as voltage drop across a 50 Ω series resistor via the oscilloscope. With this setup, a (E_r, E)-grid of 60×60 measurement points was realized. A numerically more stable data analysis was implemented in MATLAB (version 2011b) similar to what was described earlier by Pike et al.[186]. The polarization data $P_{FORC}^-(E_r, E)$ was smoothed on a 5×5 grid before fitting a polynomial surface of second order to each data point (E_r, E). The second-order polynomial fit allowed an easy extraction of the switching density as mixed second derivative according to

$$\frac{\partial^2 P_{FORC}^-(E_r, E)}{\partial E_r \partial E} = \frac{\partial^2 \left(a_0 + a_1 E_r + a_2 E + a_3 E_r^2 + a_4 E^2 - a_5 E_r E\right)}{\partial E_r \partial E} = -a_5, \qquad (3.12)$$

where $a_0, a_1, \ldots a_5$ are constants.

3.7 Harmonic Analysis

Morozov and Damjanovic utilized a sinusoidal field excitation and a lock-in amplifier for a harmonic analysis of the polarization response to study deaging (or wake-up) behavior. They found phase jumps of 180° in the third harmonic at the transition from a pinched to an open hysteresis during field cycling treatment. Extracting the time or number of cycles at which this transition takes place for different temperatures, they determined activation energies for the underlying processes from an Arrhenius plot. A phase jump marks a clear transition point, whereas e.g. a complete or incomplete merging of peaks is hard to define.

A modified approach was developed for this work[70]: Instead of the lock-in technique, a numerical Fourier series expansion was applied. This allows to obtain all harmonics in one stretch. The direct measurement would need to be repeated for each of the nine harmonics with one lock-in amplifier channel or nine lock-in amplifier channels have to be available. Moreover, the complete response of transient currents and polarization versus electric field is recorded because nothing but a normal dynamic hysteresis measurement is performed—with sinusoidal instead of triangular excitation. As described in section 3.2, a TF Analyzer 3000 by aixACCT Systems was used to measure the P-E hysteresis three times per order of magnitude in logarithmically equidistant steps. A software upgrade was developed by Bernd Reichenberg at aixACCT Systems to enable P-E measurements also between the initial measurement and 1 s of cycling treatment, which is usually not possible.[3] This modification widened the accessible time scale and with it the suitable temperature range for the intended study of the field cycling kinetics. Other limitations are given by dielectric breakdown at higher temperatures before wake-up is completed or the onset of fatigue could be observed or unsuitably long measurement times at lower temperatures. Thus, a preliminary series of experiments was necessary before the suitable temperature range and steps could be chosen.

---

[3]The standard "Fatigue Measurement" environment[189] measures an initial P-E curve after zero cycles and the next curve just after 1 s. Thus, for a frequency of e.g. 100 Hz or 1 MHz, there is no data available within the first 100 or 10^6 cycles, respectively.[189]

3.7 Harmonic Analysis

Fig. 3.14 illustrates the different steps of this novel approach of harmonic analysis to extract activation energies:

1. Perform dynamic hysteresis measurements at increasing numbers of switching cycles within the "Fatigue Measurement" environment.
2. Numerically calculate the Fourier series expansion of each P-E measurement.
3. Extract the phase jumps, i.e. the transition points for pinched $\rightarrow$ open hysteresis, for certain harmonics.
4. Plot the transition point versus 1/temperature (Arrhenius plot).
5. Repeat 1) to 4) at different temperatures.

In this work, the fatigue measurements were recorded for 100 Hz, 1 kHz and 10 kHz at each temperature to check for a frequency dependence.

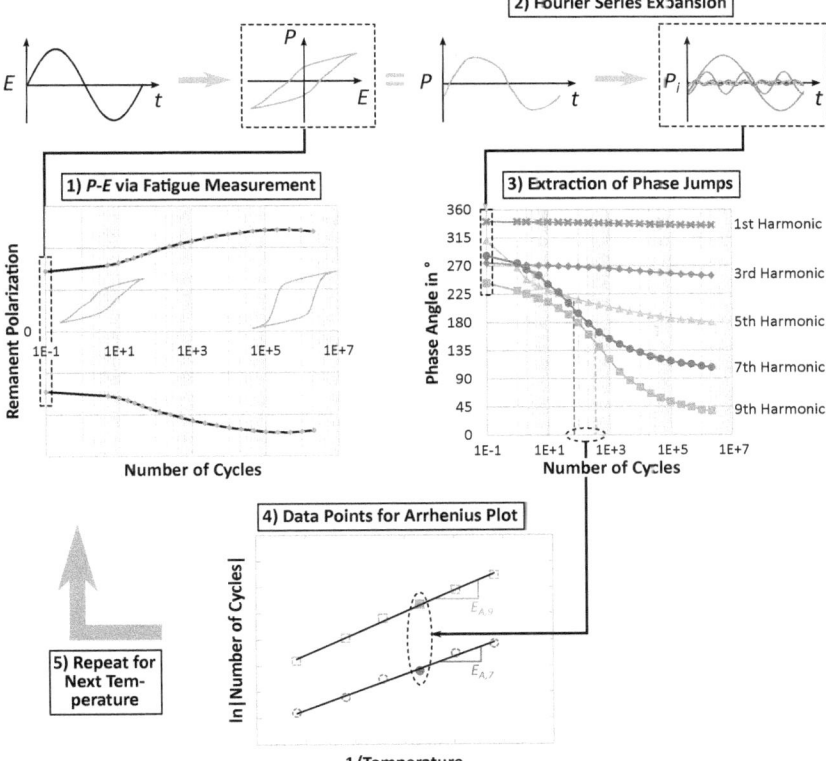

Fig. 3.14: Sketch of the measurement procedure to obtain activation energies from the modified harmonic analysis approach used in this work. The steps 1) to 5) are explained in the text above.[70]

Fig. 3.15 illustrates why the phase of the odd harmonics determine the hysteresis constriction or opening: The impact of a 180°-phase flip of the third harmonic on a Preisach model hysteresis with uniform switching density. Ideally, even harmonics exhibit an amplitude of zero because they do not fulfill the half wave symmetry required by a hysteresis symmetric to the origin of the P-E graph.[38, 190]

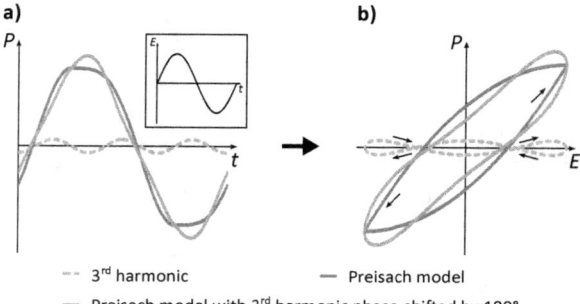

Fig. 3.15: Role of the third harmonic in case of a Preisach model hysteresis (most simple case: uniform switching density): a) modelled polarization response vs. time to a sinusoidal excitation for the third third harmonic (dashed grey line) obeying the Preisach model (red solid) and with a phase shift of 180° (blue solid line), b) resulting P-E unpinched and pinched hystereses.[48]

4 Stabilization of the Ferroelectric Phase in Hafnium Oxide

Ferroelectricity is a rather universal behavior in HfO_2 thin films. Ferroelectricity was induced by **different deposition methods** including

- atomic layer deposition (ALD)[1, 132, 138, 140, 141, 142, 139]
- chemical vapor deposition (CVD)[191]
- sputter deposition[35]
- sputter deposition[35]
- pulsed laser deposition (PLD)[192, 193]
- chemical solution deposition (CSD)[194, 195]

As discussed in section 2.1, an important fact to prove ferroelectricity is the switching between two stable states by an electric field. The sample needs to be able to withstand this field without suffering dielectric breakdown—the optimization of a deposition process to guarantee this film quality is a crucial prerequisite. **Various dopants** have been used to achieve ferroelectric polarization hystereses—most prominent examples are Si, as the dopant used in the original publications by Boescke[1, 196, 149] and the lanthanoids[197, 37]. CSD has recently been proven as a convenient way of examining a wealth of dopants, by extending the formerly known pool of suitable dopants (see below and also Table 4.1) by Nd, Sm, Er and Yb concluding that all lanthanoids are anticipated as suitable stabilizers for FE HfO_2 thin films.[195] Moreover, FE in a binary mixture with ZrO_2[142] and even ferroelectric behavior in pure HfO_2[198] were reported. Besides TiN as the original material, **different electrodes**, such as Pt[199, 194], RuO_2[200], Ir[201, 202, 191] and—as will be shown in section 4.5—TaN[172, 203], have been used to establish ferroelectric properties in HfO_2/ZrO_2 based films. The main restriction of the "universality" claimed here is given by the fact, that up to now, the observation of ferroelectricity is limited to thin films. This, on the one hand, has to be understood and on the other hand poses a challenge for both electrical and structural analysis.

In the following, a brief review of earlier work in combination with recent simulation results will be given. Subsequently, experimental results about the role of dopants based on a comparison of Si and Sr, film thickness, anneals and electrode materials based on $Gd:HfO_2$ are discussed to deepen the understanding of stabilization mechanisms in the MIM film stacks. Sections 4.4 and 4.5 are based on results obtained and published by Michael Hoffmann within his work as a student assistant supervised by the author of this thesis.

4.1 Previous Work and Simulation Results

In general, high symmetry phases are favored by higher temperature, high pressure and particle size reduction and disorder induced by dopants and defects like vacancies. A temperature increase equals an increase in phonon energy (lattice vibrations). In higher symmetry phases, this disorder is anticipated to results in a faster increase of entropy, which makes them more favorable.[61] High pressure compresses the unit cells increasing the coordination number of the atoms. Higher coordination numbers can be adopted more easily in high-symmetry structures.[204, 205] Surface energy is negligible in bulk ceramics with grains sizes of $> 1\,\mu m$, but gains increasing importance for grain sizes < 100 nm.[206] Depending on the specific values of surface energy, the phase with lowest surface energy contribution will become stable as the particle size decreases.[207] Some general trends have been reported claiming a stabilization of higher symmetry phases at reduced dimensions.[4] Moreover, it can be argued that the surface atoms have a lower coordination number[208] and thus, a higher symmetry phase would mitigate the overall loss in coordination. Similar arguments have been applied to explain the general role of dopants, vacancies and other defects: Simpler (i.e. higher symmetry) unit cell structures can more easily accommodate disorder[209, 210, 211]. This is similar to what has been explained for size reduction and loss in coordination. However, in practice, the situation is rather complex and the type of dopant as well as its location in the respective unit cell play a role—especially in complex unit cells like the perovskite cell of conventional ferroelectrics.[204, 205, 212, 213, 214, 215, 216, 161]

For hafnia, the stable bulk phase is a monoclinic polymorph, i.e. a lower symmetry phase. Consequently, the above-mentioned effects can be used to shift the phase stability towards higher symmetry phases such as orthorhombic, tetragonal or cubic polymorphs of hafnia.[207, 217] For the sister oxide, ZrO_2, which became the state-of-the-art high-k material in the capacitors of dynamic random access memories (DRAMs) at about the same time, the tetragonal phase becomes stable already for films with thicknesses below 10 nm. For hafnia, thicknesses below 2 nm are required, which on the contrary increase the crystallization temperature.[149]

A literature overview about the stabilizing effect of tri- and tetravalent dopants was given in previous dissertations.[149, 150]. The interested reader is especially referred to Tabelle (Table) 3.1 in J. Müller's dissertation[150]. A translation by the author of the present work can be found in Appendix A together with the respective references. For reasons of charge neutrality, trivalent dopants induce one oxygen vacancy per two dopant atoms incorporated.[218, 219, 220]. Besides the valence of the dopant, also the dopant size plays an important role. The created oxygen vacancies only tend to be located next to the dopant if it is smaller than the host cations (Hf or Zr). Else they are found adjacent to Hf or Zr.[220, 221] Moreover, it should be considered that e.g. Si and Al typically exhibit a four- to sixfold coordination in oxides and are not expected to be able of seven- or eightfold coordination.[211, 221, 220, 222, 223, 224] Dopant size was also found to play the dominating

---

[4]Against former understanding, it has been argued that the effect of a size reduction in ionic systems is similar to the application of a negative pressure, which expands the unit cell and increases ionicity of the bonds. According to the previous argument on the effect of pressure favoring higher-symmetry phases, these phases should be disfavored by size reduction. However, the cell expansion was explained to results in more symmetric lattice parameters and thus in higher symmetry. The situation is different in strongly covalent systems such as Si or Ge crystals because the ionicity induced by cell expansion is less significant.[206]

role in the impact on phase stability in thin films—at least for the systems studied up to now.[150, 197, 37]. The structural characterization was mainly done by GIXRD due to the thickness of only around 10 nm[1, 196, 142] (or 16 nm for Al[138]). As argued by J. Müller, strong hints exist, but clear phase assignments are hampered by the low thickness, phase mixtures and the anticipation of texture and stress.[150] The recent understanding includes the following phase transitions from lower to higher symmetry HfO_2 polymorphs depending on the amount of dopant species incorporated (from low to high content)[150, 197, 37].

- **smaller dopants**, Si^{4+}, Al^{3+}, or the equally sized Zr^{4+}:
  monoclinic $P2_1/c$ → orthorhombic $Pca2_1$ → tetragonal $P4_2/nmc$ (→ cubic $Fm3m$ for Al^{3+})
- **larger dopants**, Y^{3+}, Gd^{3+}, La^{3+}, Sr^{2+}:
  monoclinic $P2_1/c$ → orthorhombic $Pca2_1$ → cubic $Fm3m$

A closer look on Fig. 2.11 reveals that the fractional coordinates (position relative to the lattice vectors) of Hf ions in the tetragonal and cubic cell are identical. They only differ in the fractional coordinates of the O ions, which in the cubic cell are precisely in the middle of the surrounding Hf ions (fractional coordinates along a, b, c directions are either 0.25 or 0.75). In (GI)XRD, the contribution of the O ions is negligible and thus a differentiation between a tetragonal and a cubic cell becomes difficult for a/c-ratios close to 1. J. Müller showed corresponding XRD simulations in Abbildung (Figure) 4.6 of his thesis[150]: The only difference between the cubic and the tetragonal cell with an a/c ratio of one is a small peak around $2\theta = 43°$, which stems from the difference in O positions. There are two problems that complicate relying on this peak for phase differentiation: From a experimental point of view, this peak position is unfortunate since it coincides with a peak of TiN as the standard electrode material. From a symmetry point of view, it seems unlikely that the fractional coordinates stay constant as the surrounding Hf ions approach their positions in the cubic cell. A tetragonal cell ($a = b = 5.04$ Å, $c = 5.14$ Å) contracted toward a more cubic shape ($a = b = c = 5.04$ Å) is anticipated to exhibit a smooth transition of its O positions toward those of the cubic cell. However, the given reference to a paper by Fujimori et al.[225]) and references within that paper supports O positions deviating from the one in the cubic cell despite cubic lattice parameters (pseudo-symmetry). This could be attributed to asymmetry induced by the high amounts of admixture of either yttria (up to 18 cat%)[225] or ceria (50 cat% and more)[226]. Consequently, in this work, "cubic" and "tetragonal" refers to the absence or presence of tetragonallity in the unit cell parameters rather than to the precise positions of the O ions in the cell as they are not accessible with the standard methods applied here.

Tab. 4.1: Ionic radii in pm of Hf and different dopant elements in different coordination numbers in a crystalline lattice.[227, 228] The lines in bold font indicate the relevant coordination numbers in the respective phases indicated by the superscripts: M – monoclinic P2$_1$/c, O – orthorhombic Pca2$_1$, T – tetragonal P4$_2$/nmc, C – cubic Fm3m

smaller ← | → larger

Coordination Number	Si^{4+}	Al^{3+}	Hf^{4+}	Zr^{4+}	Y^{3+}	Gd^{3+}	La^{3+}	Sr^{2+}
IV	40	53	72	73	–	–	–	–
V	–	62	–	80	–	–	–	–
VI	54	67.5	85	86	104	107.8	117.2	132
VII^{M,O}	**61**^{a}	**75**^{a}	**90**	**92**	**110**	**114**	**124**	**135**
VIII^{O,T,C}	**68**^{a}	**83**^{a}	**97**	**98**	**115.9**	**119.3**	**130**	**140**
IX	–	–	–	103	121.5	124.7	135.6	145
X	–	–	–	–	–	–	141	150
XI	–	–	–	–	–	–	–	–
XII	–	–	–	–	–	–	150	158

^{a}Obtained from a linear extrapolation of existing data as Fig. 7.10 in the book by Rohrer [229] suggests.

Here, "smaller" and "larger" are defined by a comparison of the substituent's ionic radius as compared to Hf^{4+} in VIII-fold coordination. A recent simulation study also concluded that the dopant size is the predominant parameter to decide between a transition to tetragonal or directly to cubic with increasing dopant content.[230] Table 4.1 gives an overview about the Shannon radii of each dopant for different coordination numbers. Two coordination numbers are of interest for the respective HfO$_2$ phases: VII (monoclinic, orthorhombic), VIII (orthorhombic, tetragonal, cubic). Radii of Si and Al are not tabulated for these coordination numbers. The electron configurations for both these elements only exhibit occupied orbitals up to 2s and 2p orbitals, instead of 3d, 4d, 5d and 4f like for the larger dopants. Preferred coordinations of Si and Al in oxides are IV-fold (tetrahedral) to VI-fold (octahedral) as mentioned above.[211, 221, 220, 222, 223, 224] From that perspective, it would not be surprising if the incorporation of these elements results in additional effects. Recently, the group around Jacob Jones at North Carolina State University started to reproduce experiments and validate claims regarding the stabilization of different hafnia polymorphs at certain pressure, stress and temperature conditions. For Si doping, their findings included:

- Si can be substituted for Hf into the monoclinic P2$_1$/c lattice up to between 5 cat% and 9 cat%, which is in contrast to the equilibrium phase diagram. This phase diagram just shows a solid solution of pure HfO$_2$ and an HfSiO$_4$ phase and no solubility of Si in HfO$_2$. This substitution deforms the monoclinic lattice towards the cubic Fm3m cell.[231] For thin films, Böscke reported a solubility of up to 14 cat%.[149]
- An expansion is observed for the incorporation of the smaller Si on Hf lattice places in the monoclinic unit cell, which is opposed to a simple argument of dense packaging.

4.1 Previous Work and Simulation Results

Explanations include: 1) a small amount of Si interstitials; 2) the preferred VI-fold coordination of Si instead of a VII or VIII-fold coordination of Hf, which leads to oxygen vacancies contracting the cell; 3) different electronegativities of Hf and Si resulting in different ionicity of the Hf-O and Si-O bonds and a cell deformation, 4) the occurrence of a non-equilibrium Si:HfO$_2$ phase altering the vacancy or interstitial profile of the remaining monoclinic cells in a contracting manner.[232]

- Pressure-induced phase transitions for pure and Si doped HfO$_2$ powders were studied. A monoclinic → orthorhombic phase transition occurred for increasing pressures. Both the FE Pca2$_1$ and the AFE Pbca phase could explain the measured diffractograms and a differentiation was not possible. Since, the transition pressures did not strongly vary for the different concentration, it was concluded, that the Si doping alone cannot explain the field-induced phase transition observed by Böscke et al.[1, 196] and other thin film studies.[233]

- A metastable Si:HfO$_2$ phase occurred for different calcination temperatures and dependent on the particle sizes of the source powders.[234]

However, it should be stressed that the applicability of these findings in ceramics on polycrystalline thin films is limited. In thin films, stress/strain and surface energy effects gain importance. However, the results of both thin films and ceramics should be considered and crosschecked in future studies. The current state of research does not allow for solid conclusions. First simulation studies on the topic of ferroelectricity in hafnia and zirconia exist.[235, 236, 160, 61] Their compliant as well as their conflicting results a are discussed in the following paragraphs to form a best possible overall picture. Nonetheless, it should be kept in mind that these studies represent just the beginning of an, as it turned out, rather challenging path to explain ferroelectricity in hafnia and zirconia.

A central question is, how to stabilize the orthorhombic phase. Materlik et al.[61] performed calculations for pure HfO$_2$, Hf$_{0.5}$Zr$_{0.5}$O$_2$ and pure ZrO$_2$. For none of the three compounds, a temperature window with a stable Pca2$_1$ phase was found. At 300 K, the energy difference to the monoclinic bulk phase is 60 meV, 50 meV and 35 meV for HfO$_2$, Hf$_{0.5}$Zr$_{0.5}$O$_2$ and ZrO$_2$, respectively. Before discussing potential knobs to stabilize the FE phase considered in literature, some central properties of the ferroelectric phase and the way it can be switched from one to the other polarization state are explained.

Slight differences can be observed between the spontaneous polarizations calculated for the polar Pca2$_1$ phase: Reyes-Lillo et al.[236] found 58 µC/cm^2 and 60 µC/cm^2 for pure ZrO$_2$ and pure HfO$_2$, respectively. Materlik et al.[61], who carefully adjusted their pseudo-potentials to be capable of reproducing experimentally found values for bulk phase transitions and phonon modes reported for ceramics, calculated the same value for zirconia, while for hafnia they only obtained 50 µC/cm^2. Huan et al.[160] found 52 µC/cm^2 for hafnia, which is in good agreement with that value, while values by Clima et al.[235] show a rather wide spread from 41 µC/cm^2 to 53 µC/cm^2. This spread results from a different choice of boundary conditions. Clima et al.[235] fixed the unit cell parameters to values estimated from peak positions in GIXRD measurements of ferroelectric thin films. Thus, the uncertainty in these estimates directly affects the calculated P_s-values. In addition to the Pca2$_1$-phase that Böscke et al. proposed to explain the FE properties of their films, Huan et al.[160] reported another polar phase that could account for the observed electrical behavior. This phase is of space group Pmn2$_1$ and shows a P_s of 56 µC/cm^2. However, to the authors knowledge, this phase was neither observed in any TEM study nor does its XRD pattern fit to the experimentally observed diffractograms as will be explained in section 4.2.

Clima et al.[235], who provided the first study explicitly related to ferroelectricity in hafnia did not focus on the stability of the ferroelectric phase. They just considered the two polarization states in the Pca2$_1$ phase and the transition between the two. Switching barriers of 260 ... 630 meV/f.u. (f.u. = formula units) were calculated depending on the unit cell parameters. The corresponding double-well potential can be fitted by a fourth-order polynomial, which would correspond to an second-order phase transition (see section 2.2) and the unpolar transition state (O ions at the cusp of the switching barrier) passed during switching looked similar to an orthogonalized monoclinic unit cell. These results are contradicted by the study of Reyes-Lillo et al.[236], who found the tetragonal parent phase to represent the local energy minimum with no unstable modes. Therefore, a first-order transition is expected, although a second-order transition would be permitted by symmetry analysis. The energy barrier between the tetragonal P4$_2$/nmc and the orthorhombic Pca2$_1$ in ZrO$_2$ was calculated to be 35 meV/f.u. and their energetic difference is nearly zero. Thus, the transition path used by Clima et al. might not have been the energetically most favorable, which automatically explains the orders of magnitude higher barriers. Huan et al. also considered the tetragonal P4$_2$/nmc phase the parent phase of the ferroelectric phase. By deforming it along its [110] direction, a barrier of 6 meV/f.u. is passed before the cell relaxes into the FE Pca2$_1$ phase. For the Pca2$_1$ phase, the deformation direction equals its polar [001] direction.

The effect of strain was also considered by other groups to explain the stabilization of the FE phase. Reyes-Lillo et al.[236] argue that reasonable levels ($\approx$1 %) of compressive epitaxial strain normal to the lattice vector of a cubic ZrO$_2$ parent phase can stabilize the orthorhombic phase compared to the tetragonal one. A substitution of Zr by Hf is suggested to stabilize a ferroelectric phase. In pure HfO$_2$, the orthorhombic phase is favorable compared to the tetragonal by 23 meV/f.u. Therefore, a much wider range of epitaxial strain is anticipated to allow for a stable Pca2$_1$ phase if—as for ZrO$_2$—the monoclinic phase is suppressed. However, it should be considered that their calculated lattice parameters and experimental data shows only a good agreement for ZrO$_2$, but not for HfO$_2$. The latter values are systematically smaller for all phases, by several percent, which might hint at a problem with the Hf pseudopotentials used. This also affects the outcome of strain considerations. Clima et al.[235] found that in both [001] and [110] direction of the FE Pca2$_1$ phase, an uniaxial compressive/tensile stress causes a barrier decrease/increase, respectively. According to calculations by Materlik

4.1 Previous Work and Simulation Results

et al.[61], no reasonable in-plane strain conditions were found that would make the $Pca2_1$ phase the energetically most favorable phase if the polar c-axis in an out-of-plane direction as it is required by the electrical results for thin film capacitors.

Simulations by Materlik et al.[61] suggest that surface energy might be dominant and other effects of minor impact as will be shown later. Surface energy was modeled by linearly interpolating between the existing values for 1) the tetragonal and the monoclinic phase to obtain the surface energies for the orthorhombic phase as well as 2) between those of ZrO_2 and HfO_2 to estimate the surface energies for $Hf_{0.5}Zr_{0.5}O_2$. Cylindrical grains with their diameter equal to their height were assumed. A fixed height of 9 nm was used to successfully reproduce the results by J. Müller et al.[142] obtained for 9 nm thick films: At 300 K, the stable phase in HfO_2, $Hf_{0.5}Zr_{0.5}O_2$ and ZrO_2 are monoclinic $P2_1/c$, ferroelectric orthorhombic $Pca2_1$ and tetragonal $P4_2/nmc$, respectively. With T fixed to 300 K, grain size/film thickness windows for a stable FE phase were derived to be 3...5 nm for HfO_2 and 8...16 nm for $Hf_{0.5}Zr_{0.5}O_2$, respectively. For ZrO_2 no such window existed. The role of surface energy is supported by recent results of upt to 400 nm thick FE films of pure zirconia with columnar grains that conserve the impact of surface energy also for thicknesses far beyond the typical scale of 10 nm.[237]

Finally, the impact of electric fields was considered by Materlik et al.[61]. A electric field of 1 MV/cm reduces the energy of ZrO_2 by about 10 meV/f.u., which can explain the field-induced phase transition. In bulk ZrO_2 about 3.5 MV/cm are required at 300 K, whereas for 9 nm thickness, the values of 1 MV/cm presented by J. Müller et al. is sufficient.

Fig. 4.1 is based on calculations by Materlik et al. and summarizes the estimated impact of the different knobs available to alter the phase stability in favor of a room temperature stable FE $Pca2_1$ phase. Routes to form a FE phase were exemplarily suggested by Huan et al.[160]. First the $P4_2/nmc$ parent phase needs to be stabilized. $T = 2200$ K, at ambient pressure or $T = 1800$ K at 10 GPa would be suitable conditions. Next, internal or external fields or other effects could be applied to achieve one of the polar phases before quenching the sample to ambient conditions. This route it not compatible with processing schemes required in CMOS fabrication, but it might be feasible for ceramics.

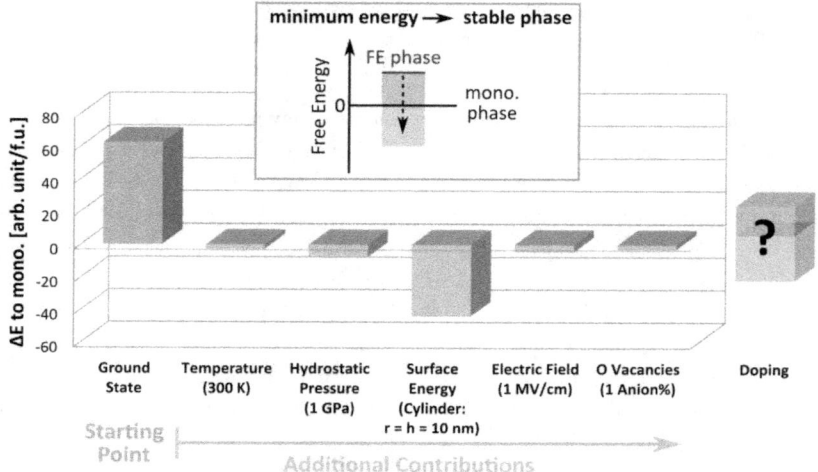

Fig. 4.1: Overview of the impact of different parameters on phase stability as calculated by Materlik et al.[61] using the example of $Hf_{1-x}Zr_xO_2$. ΔE is the difference in free energy between ferroelectric and monoclinic phase originating from the respective contributions.[93]

4.2 Proof of the Ferroelectric Phase

Already in the first publications on FE HfO_2, the $Pca2_1$ phase (in the setting $Pbc2_1$) has been proposed to explain the ferroelectric behavior observed in the P-E, C-V and macroscopic piezoelectric measurements. Lots of considerable hints toward a ferroelectric phase existed, namely:

- Butterfly-shaped small-signal C-E curves and polarization hysteresis were reported.[132, 238]
- These features are found in the transition region between between monoclinic and tetragonal/cubic phases and coincide with a k-value change.[138, 132]
- GIXRD and grazing angle attenuated total reflection Fourier transform infrared spectroscopy (GATR-FTIR) hint at a third phase in this transition region.[132, 138, 139]
- The samples show subloop and fatigue behavior similar to conventional ferroelectric.[238, 239, 240, 241]
- They show a remarkable retention of stacks not particularly designed as charge trapping devices.[240, 242, 238]
- A ferroelectric-like piezoresponse was found in macroscopic laser interferometry[1, 238] and microscopic PFM measurements.[243]
- The threshold voltage shift in the manufactured FeFETs is opposed to what is expected for electron injection (as it was shown for exactly the same stacks with Si concentrations outside the FE dopant concentration window).[239, 244, 242]

4.2 Proof of the Ferroelectric Phase

As Fig. 4.2 by Sang et al.[171] exemplarily shows, a clear butterfly-shaped hysteresis exists in the curve of small-signal permittivity as function of electric field. The maxima in this curve coincide with the steep edges in the corresponding polarization hysteresis. Moreover, the minimum k-values (outside the peaks), which represents the dielectric properties without any domain wall contributions, is an indicator of the crystallographic phase of the material. For this sample with marked FE behavior in the electrical characteristics, the k-value is between what has been reported for a higher-symmetry cubic and a lower-symmetry monoclinic phase, which might hint toward the presence of an intermediate phase. However, a parallel and/or series arrangement of monoclinic and tetragonal/cubic grains could also cause this transition. This topic is addressed in more detail in section 5.6.

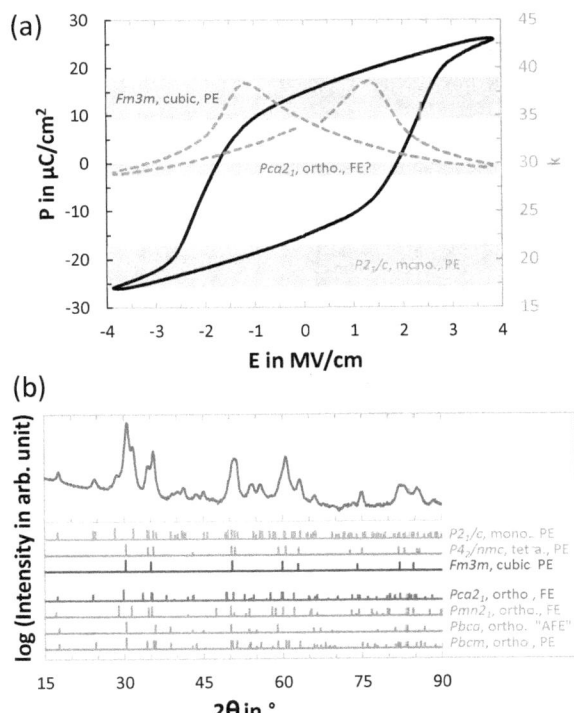

Fig. 4.2: Example outlining the difficulties in clear phase identification from GIXRD (27 nm thick Gd:HfO$_2$ film; N$_2$ anneal at 650 °C for 20 s): a) ferroelectric polarization and small-signal capacitance hysteresis (triangular field sweep with aixACCT TF Analyzer 3000) and b) corresponding GIXRD measurement curve and reference patterns for different phases from powder diffraction files.[171, 93]

However, up to that publication, the clear structural identification of the Pca2$_1$ phase was hampered by 1) the existence of a wealth of potential hafnia phases with a high number of overlapping peaks 2) the polycrystalline nature with potentially a phase mixture and 3) the low thickness of the thin films. A combination of scanning transmission electron microscopy

(STEM) and electron diffraction has been used to prove the presence of the $Pca2_1$ phase in a 27 nm thick Gd:HfO$_2$ film with TiN top and bottom electrodes anneals in N$_2$ at 650 °C for 20 s. Fig. 4.2 shows a selection of potential hafnia phases in different crystallographic projections. High-angle annular dark field (HAADF) STEM with a revolving STEM (RevSTEM) approach allows an accurate determination of lattice parameters from real-space images. Comparing these images to the unit cell sketches, the monoclinic $P2_1/c$ phase as well as the orthorhombic, non-centrosymmetric (i.e. potentially FE!) $Pmn2_1$ can be ruled out.[5] Unfortunately, all orthorhombic phases basically look the same as image contrast in the HAADF-STEM images is dominated by the Hf atoms (much higher Z than O). Mainly the Hf atoms contribute to the contrast in the image, but the major differences between these phases are the O positions.

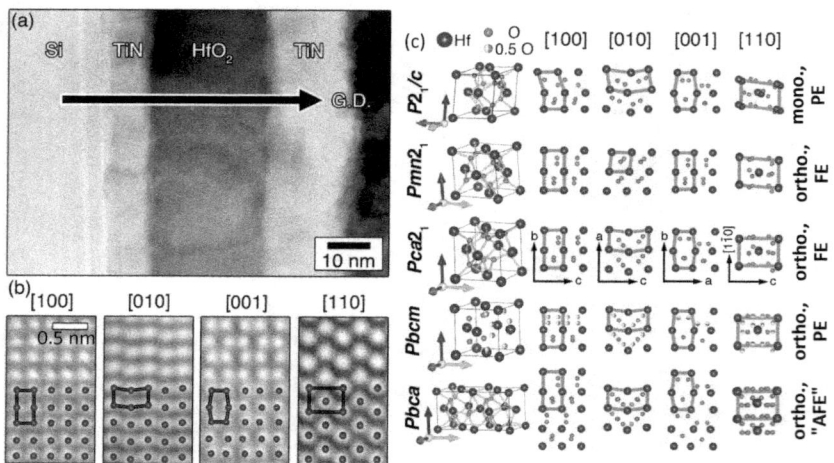

Fig. 4.3: a) Bright field STEM image showing the film microstructure. The growth direction (G. D.) is indicated by the black arrow. b) HAADF-STEM images acquired from four different grains superimposed with the Hf atom column arrangement projected along the four major zone axes, which are the same for $Pca2_1$, $Pbca$, and $Pbcm$ phases. The probe semi-convergence angle was 14 mrad for [010] and [110] images, and 20 mrad for [100] and [001] images. c) The crystal structure of five HfO$_2$ phases projected along four major zone axes. Lattice vectors settings: vector a is red, vector b is green, and vector c is blue.[171]

In order to verify if there is an asymmetry along the [001] axis of the grain, which is essential for the non-polar ferroelectric $Pca2_1$ phase, position averaged convergent beam electron diffraction (PACBED) has been applied. The measured image shows a marked asymmetry coinciding with the simulated pattern for the $Pca2_1$ phase. The two remaining centrosymmetric orthorhombic phases, $Pbca$ and $Pbcm$ can be ruled out and only leaves one explanation for these solely structural results: The originally claimed $Pca2_1$ phase is present in a sample which is known to exhibit FE-like properties similar to the ones observed in the past. Together with clearly tetragonal/cubic high-temperature data presented for similar thin films[150, 149],

---
[5]Cubic and tetragonal phases were already ruled out because of the lattice parameters and thus, not considered any further.

this evidence implies, that Gd:HfO$_2$ is rather a displacive instead of an order-disorder ferroelectric. In order-disorder ferroelectrics, the dipole of the unit cell does not vanish, but all dipoles point towards random directions in the higher temperature state.[22] It is expected that this phase is also the source of ferroelectricity in other HfO$_2$ and ZrO$_2$-based samples.

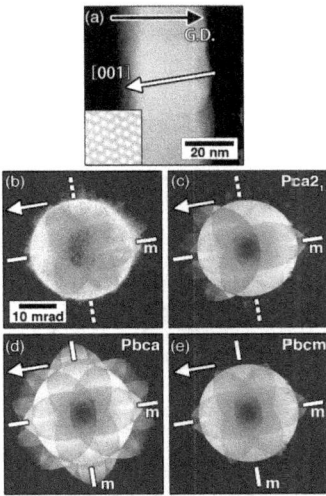

Fig. 4.4: a) STEM image of the Gd:HfO$_2$ thin film with a 2 × 2 nm^2 inset showing the projected structure ([110]) corresponding to the experimental PACBED pattern in b). The sample thickness is 10 nm, as determined by PACBED. The growth direction (G.D.) is indicated by the black arrow. Simulated PACBED patterns for the c) polar orthorhombic Pca2$_1$, d) non-polar orthorhombic Pbca, and e) non-polar orthorhombic Pbcm phases. Sample thickness was simulated to match the experiment. Solid and dashed bars indicate the presence or absence of mirror symmetry, respectively.[171]

4.3 Impact of Smaller and Larger Dopants

As mentioned in section 4.1, differences between the effect of dopants with smaller and larger ionic radii than Hf were reported already in the first publications[197, 37]. First of all, it needs to be mentioned that dopant incorporation usually increases the crystallization temperature as shown in Fig. 4.5.[149] Therefore, a comparison of films with same thickness and electrodes subjected to the same annealing conditions can never be 100 % fair. In this subchapter, basic structural and electrical findings for HfO$_2$ doped with Si, as a representative of the smaller dopants, are contrasted with films doped with Sr, an examples of the larger dopants. Si is the element used in the original publications by Böscke et al.[1] and Sr is an alkaline earth metal—bivalent and larger than Hf. The influence of process parameters and electrodes is discussed separately in sections 4.4 and 4.5 using the example of HfO$_2$ doped with Gd, a larger trivalent dopant of the lanthanoid series.

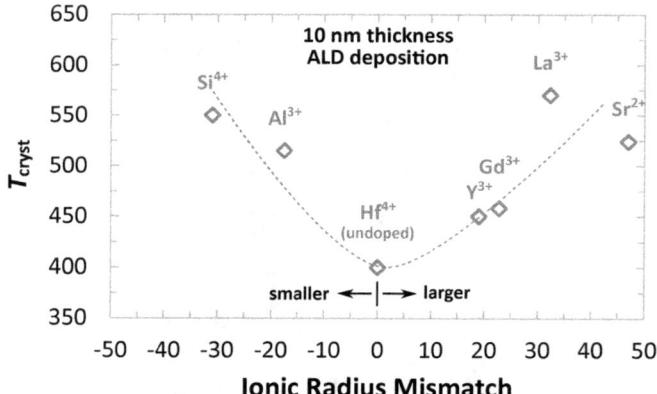

Fig. 4.5: Crystallization temperature for 10 nm thick FE HfO_2 films with different dopants as obtained from temperature dependent XRD measurements. Data from J. Müller[150] (pure HfO_2, $Si:HfO_2$ and $Y:HfO_2$) as well as measurements performed at Fraunhofer IPMS, Dresden ($La:HfO_2$) or Imec, Leuven and Ghent University ($Al:HfO_2$, $Gd:HfO_2$, $Sr:HfO_2$).

Effect of Si Incorporation

The first part of this section is devoted to the description of structural and electrical characteristics of different Si concentrations. Like Al and Zr, Si is expected to stabilize a tetragonal phase for higher dopant concentrations. Compared to the binary mixture $Hf_{1-x}Zr_xO_2$, $Si:HfO_2$ exhibits a narrower concentration range for the transition from monoclinic $P2_1/c \rightarrow$ orthorhombic $Pca2_1 \rightarrow$ tetragonal $P4_2/nmc$. Therefore, smallest possible concentration steps were chosen around the transition region to assess the electrical and structural changes they induce. Fig. 4.6 a) shows the GIXRD scans for different ALD cycle ratios and the reference patterns from ICDD (see Tab. 3.2). For a cycle ratio of 46:1 ($HfO_2:SiO_2$ cycles), the measurement shows significant signs of the monoclinic phase. Going to lower cycles ratios of 26:1, i.e. increasing Si concentration, the peaks around 18°, 25°, 29° and 32° lose intensity compared to a rising peak around 31°. The feature at 35° changes from a broadened peak, obviously consisting of multiple peaks, to a sharper peak with only one shoulder at its lower 2θ side. These features coincide with what is expected from decreasing monoclinic and increasing orthorhombic or tetragonal phase fractions. Upon further increasing the Si concentration, the described trends continue. At a cycle ratio of 18:1, the peaks at 18°, 25°, 29° and 32° are not observed. The peak around 31° still possesses a shoulder at its higher 2θ side and also the peak signature between 80° and 90° evolved from a noisy pattern with a peak around 89° towards a broadened two or three-peak structure between 83° and 86°. Towards a cycle ratio of 12:1, the pattern more and more matches the expectations from a purely tetragonal film.

4.3 Impact of Smaller and Larger Dopants

Clear distinctions from a dominating orthorhombic or cubic phase can be made:

- one clear peak at 31°, two around 35°
- a weak one rising at 43°
- the feature around 52° clearly stemming from more than just a single peak
- three peaks around 61°
- two maxima around 75°
- more than just two distinct peaks around 84°.

Whereas the identification of dominating monoclinic and tetragonal phase fractions at lowest and highest Si concentrations is convincing, a precise determination of orthorhombic phase fraction is not possible without carefully refining the patterns. As stated before, the concentration range for the phase transition is rather small compared to Zr incorporation. Assuming a certain degree of inhomogeneity within the film (grain size, local dopant/defect concentration) and towards the electrodes (O pulled out by oxidizing TiN to Ti-O-N), a multiphase coexistence throughout all steps between highest and lowest concentration would not be surprising.

Fig. 4.6 b) shows a selection of P-E hystereses and transient I-E curves as the electrical counterparts to the structural data from Fig. 4.6 a). All curves are recorded from pristine (uncycled) samples. For an ALD cycle ratio of 46:1 (HfO_2:SiO_2 cycles), a purely paraelectric response is identified. Via 32:1 to 26:1 a ferroelectric hysteresis opens. A main switching peak occurs at 1 MV/cm for both concentrations. For the 26:1 case, the hysteresis is slightly pinched, because of a shoulder at the lower field side of the main switching peak and small and broad second switching peak beyond 2.5 MV/cm. The current level at fields beyond the main switching peak is markedly higher than before the peak. Since the curves do not show significant leakage contributions, this might hint at a phase change or other non-linear dielectric behavior (compare section 2.4). However, this phase or non-linearity should also be present when decreasing the field again, which does not seem to be the case. Looking at the curve of 24:1, the next higher concentration, the hysteresis becomes more pinched, but the maximum polarization remains nearly unchanged. The main switching (or relaxation) peak moved towards lower fields and is now observed around 0 MV/cm. At higher fields a broadened double-peak structure between 2 and 4 MV/cm is observed. Between the main switching peak around 0 MV/cm and this double-peak structure, the current level does not relax back to its value before 0 MV/cm. For 20:1 and 12:1 clearly pinched hystereses with nearly zero P_r are obtained as it is typical for antiferroelectric-like hystereses (compare Fig. 2.6 e) and corresponding explanations). The switching peaks of same polarity move further apart. The maximum polarization is reduced because the field amplitude seems insufficient to switch the same amount of unit cells to the polar state. As Fig. 4.6 c) shows, a trend in the peak position of the transient current curves is observed. Until 26:1 the main switching peak remains at a fairly constant position. For higher concentrations it seems split up into two switching peaks, which move further and further apart as the Si concentration increases.

A comparison of electrical and structural data suggests the scenario shown on the right side of Fig. 4.6 c):

- At lowest Si concentrations, only paraelectric behavior is observed, which stems from a purely monoclinic (space group $P2_1/c$) film. No polarization switching is possible.
- As the Si content increases towards ALD cycle ratios of 28:1, fractions of the ferroelectric, orthorhombic phase (space group $Pca2_1$) increase. The highest maximum polarization occurs for 26:1. Monoclinic phase fractions might be present until a cycle ratio of even 22:1 or 20:1 as evident from the GIXRD results. Structural inhomogeneities account for coexistence of multiple phases in this sensitive system.
- For ratios of 26:1 and below, the hystereses starts to pinch with the switching peaks of same polarity moving further apart and with the maximum polarization decreasing for cycle ratios lower than 24:1. At 12:1 no signs of ferroelectric phase remain in the GIXRD, instead strong evidence for a dominating tetragonal phase exists. Together with the AFE-like *P-E* and *I-E* curves, this hints at a field-induced phase transition into the FE phase. According to Landau-theory (see section 2.2), such a transition occurs for a ferroelectric with first-order transition in the temperature range around the Curie temperature T_C. As argued by Hoffmann et al.[172], an increasing concentration of (suitable) smaller dopants is equivalent to a T_C decrease. Consequently, higher fields have to be applied to bend the atomic potential in favor of the polar phase. Indeed, the tetragonal $P4_2/nmc$ phase is a parent phase of the polar $Pca2_1$ phase[160], which makes the scenario also likely from a crystal symmetry point of view. The barrier between the phases has been calculated to be 35 meV for ZrO_2[236] (see section 4.1). Surface energy has been shown capable of reducing the energetic difference between the phases drastically.[61] An electric field of 1 MV/cm accounts for roughly 10 meV.[61] With this energy reduction, a barrier reduction is also anticipated.

Similar results were obtained by Park et al.[245] for $Hf_{1-x}Zr_xO_2$. The main difference is the strong interference of different phases at the same concentration. In the following, the evolution of the FE properties of two samples is shown to support this statement.

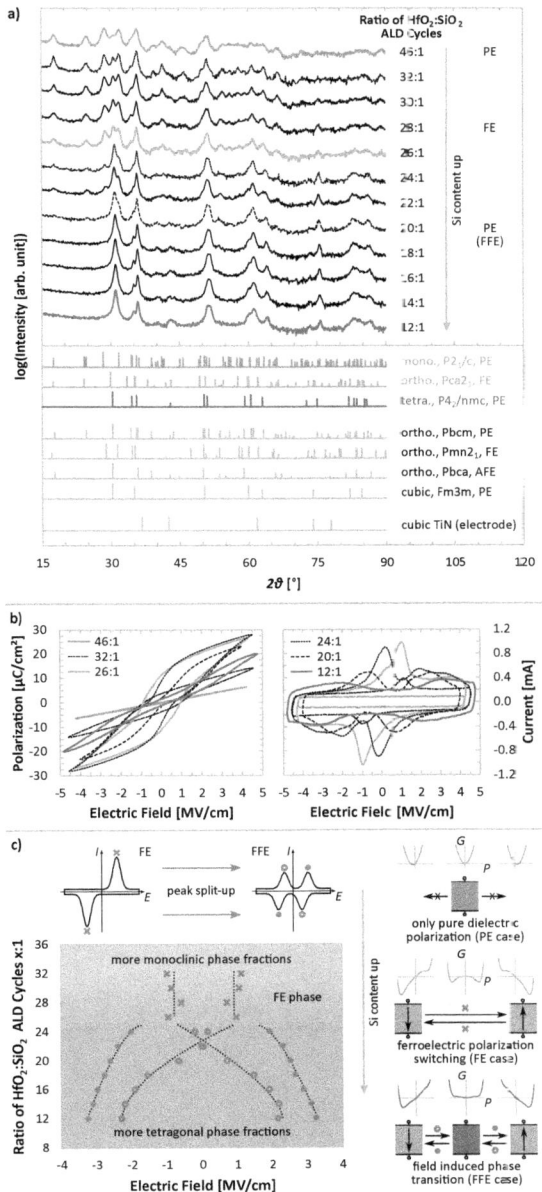

Fig. 4.6: a) GIXRD results of for 30 nm thick Si:HfO$_2$ films with respective reference patterns (see Table 3.2), b) polarization hysteresis and corresponding transient currents for selected compositions and c) Si content dependent evolution of peak position in the transient current curves and sketch of the corresponding bending of the atomic potentials for paraelectric (PE) phase, ferroelectric (FE) phase as well as for the field-induced phase transition into the FE phase (FFE). An ALD cycles ratio of 14:1 results in 5 cat% Si incorporation (XPS).

Fig. 4.7 a) and b) shows the evolution of *P-E* and *I-E* curves during field cycling for an FE (ALD cycle ration 28:1) and sample with FFE (ALD cycle ration 20:1), respectively. As summarized by Schenk et al.[48], a preconditioning step has commonly been applied to ferroelectric HfO_2 samples to open a previously constricted hysteresis. This constriction originates from internal bias fields, that typically vanish during field cycling (see. section 5.1). A similar behavior is observed for the FE sample in Fig. 4.7 a). For a purely FFE sample, internal bias fields are expected to be less prominent, since no spontaneous polarization is present during the anneal process. However, also the FFE peaks are subject to a position change and interestingly an additional FE peak arises around 1 MV/cm. It cannot be proven if the phase fractions change during this cycling sequence or if the whole evolution is solely due to the internal bias field evolution of the FE peaks and their interference with the FFE peaks. Moreover, the charges accounting for the internal bias fields of the FE regions are likely to interact with the atomic potentials of the neighboring non-polar regions. This might also give rise to an internal bias field across the non-polar phase leading to a tilt in the respective atomic potential even at zero applied field.

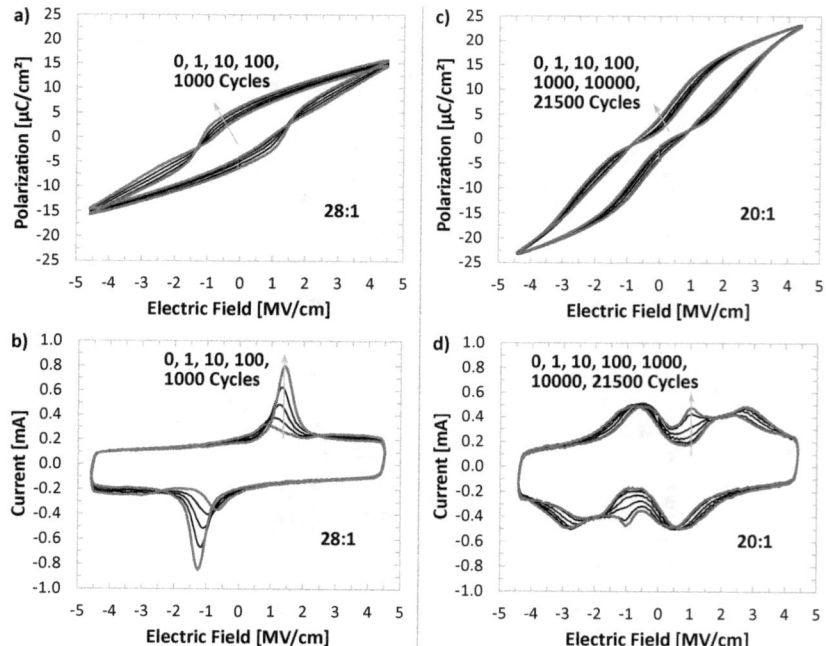

Fig. 4.7: Wake-up behavior evident from polarization hysteresis in a), c) and corresponding transient currents in b), d) for two characteristic samples: ALD cycle ratio of $HfO_2:SiO_2$ of 28:1 corresponding to an FE sample and a ratio of 20:1 corresponding to a sample at the border between FE and PE as judged by the diffractograms in Fig. 4.6.

Effect of Sr Incorporation

Compared to Si as the first dopant reported to induce ferroelectricity, Sr (similar to Y or Gd and other lanthanoids) is expected to stabilize a cubic instead of a tetragonal phase going from pure to highly doped HfO_2.[117] Fig. 4.8 shows GIXRD results for different dopant concentrations as determined by Rutherford backscattering spectroscopy. Similar to Fig. 4.6 a), the diffractograms start with dominating monoclinic features for pure HfO_2. Already for 1.2 cat% (atomic ratio of Sr/[Sr+Hf]), the peaks around 18°, 25° and especially at 29°, 32° lose whereas a peak around 31° gains intensity. Compared to the case of Si doping, the highest dopant concentrations available were not sufficient to result in a destabilization of the orthorhombic in favor of a higher symmetry phase. Thus, no proof whether this phase is a tetragonal or a cubic one can be provided for these 10 nm thin films. Moreover, a look at the ALD cycle ratios reveals what might have interfered with the nominal increase in dopant concentration: The highest concentrations of 7.0 cat% and 7.9 cat% were fabricated by applying a 30:2 and 36:3 instead of a 15:1 and 12:1 supercycle due to restrictions in the maximum number of supercycles per ALD run allowed by the tool software (e.g. $5 \times [20\ HfO_2 + 2\ SrO]$ vs. $10 \times [10\ HfO_2 + 1\ SrO]$ cycles). If the dopant distribution during annealing was not as homogenous as assumed[246], it might not have been as effective for stabilizing the higher symmetry phases.

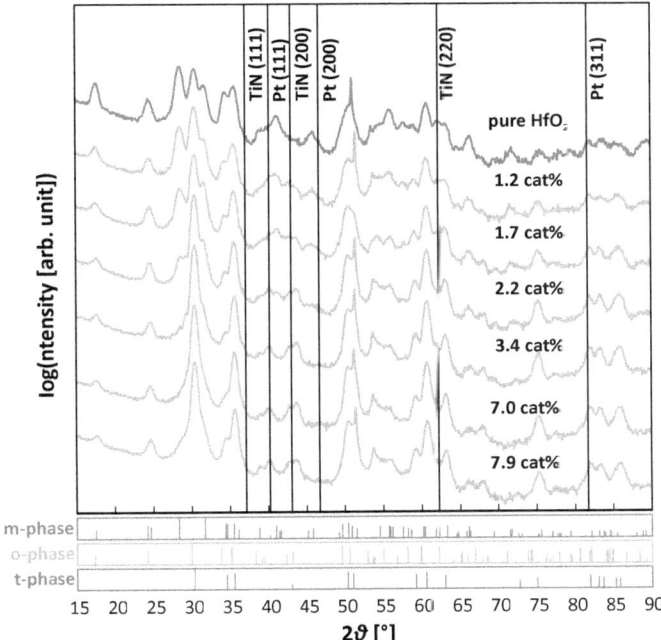

Fig. 4.8: GIXRD results of 10 nm thick Sr doped films (anneal: N_2, 800 °C, 20 s) and corresponding reference patterns (powder diffraction files see Table 3.2). This figure plots the data of ref. [141] in logarithmic form.

The corresponding electrical characteristics, remanent polarization P_r, relative permittivity k, coercive field E_c and switching field E_s are shown in Fig. 4.9 together with the hysteretic curves for three different dopant concentrations. As can be seen from Fig. 4.9 b), a P_r of $23\,\mu C/cm^2$ was achieved for 3.4 cat%. Interestingly, the coercive fields around $2\,MV/cm$ are higher than the $\approx 1\,MV/cm$ typical for Si, Al and Zr doping. Ab-inito simulation by Clima et al.[235] do not give a consistent view on what to expect with respect to dopant size. It is important to note that the calculated switching barriers are also based on fixed unit cell parameters from unrefined experimental GIXRD data. The comparability of coercive fields of different sets of samples is also limited. As explained in section 2.4, the macroscopic coercive field does not depend solely on the switching barrier itself. However, simulation results for total energy[141], i.e. the energy of formation at 0 K, agree well with two experimental observations: 1) Si is a strong stabilizer of the tetragonal phase. The energy difference between the monoclinic bulk phase and the orthorhombic phase is slightly lowered by the incorporation of 3 cat% Si, but the tetragonal phase is favored even more strongly. As temperature increases, the lower symmetry phases are energetically less and less favorable (see section 4.1). If the FE phase is less favorable than the higher symmetry tetragonal phase already at 0 K, there will be no temperature window with the FE phase being the most stable phase. 2) Doping the same amount of Sr into hafnia lowers the total energy for the tetragonal and the orthorhombic phase by a similar value, which leaves the window for a stable FE phase still open. However, the cubic phase has not been considered in these calculations.

Structural and electrical data was published by S. Mueller[139] for HfO_2 doped with Gd, which is also larger than Hf. They faced similar problems in an unambiguous assignment of the cubic or tetragonal phase. The diffractograms for highest dopant concentrations were attributed to a tetragonal or cubic phase. Also grazing angle attenuated total-reflectance Fourier transform infrared (GATR-FTIR) spectroscopy did not allow to clearly determine the phase. However, these results hint at the presence of an intermediate phase between the monoclinic and tetragonal/cubical films exactly at the Gd concentrations that induced ferroelectric characteristics in the electrical measurements. A study of different concentrations for thicker (e.g. 30 nm) films, similar to what has been presented above for Si, would be helpful. Recently, for 60 nm thick films with ferroelectric behavior, a cubic phase has been assigned to the diffractograms[193] obtained from a Bragg-Brentano geometry. In contrast, TEM studies[185] of 27 nm thick ferroelectric films suggested the presence of a tetragonal phase at the interfaces of a pristine film (as will be discussed in more detail in section 5.6). The tetragonal or cubic phase was not very prominent in the volume, but the Gd concentration of this film was not as high as in ref. [193]. A cubic phase, which is not a direct parent phase of the FE orthorhombic phase, would explain why no field-induced phase transition as for the Si doping is observed. It would also explain why the thickness range for which FE behavior has been demonstrated seems wider. As mentioned by J. Müller[150], differentiating between a cubic, a tetragonal phase with only slightly different lattice constants and a tetragonally distorted cubic phase is challenging. As mentioned above, most of the difference in the unit

4.3 Impact of Smaller and Larger Dopants

cells comes from the oxygen atoms, which barely contribute to the measurement signal in both XRD and TEM.

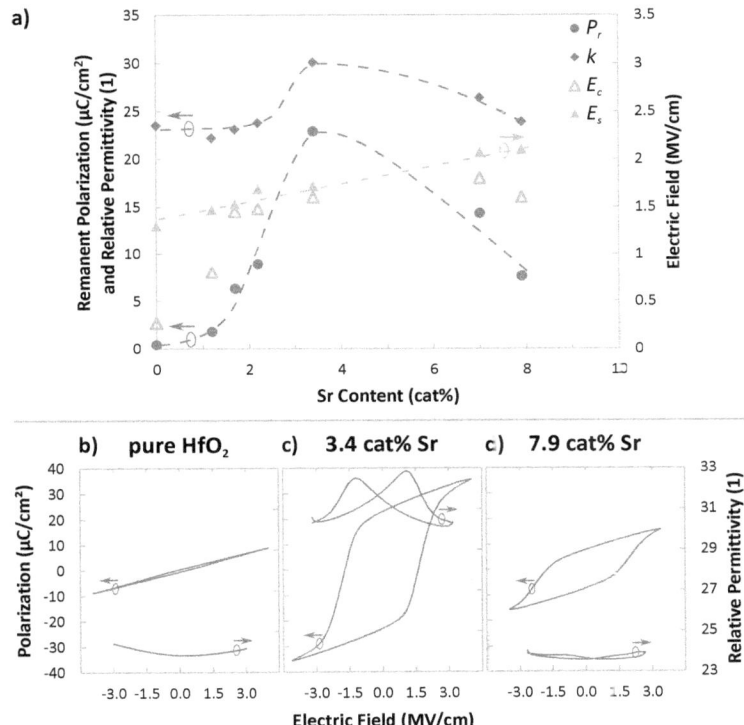

Fig. 4.9: Selection of 10 nm thick Sr:HfO$_2$ films (N$_2$ anneal at 800 °C for 20 s): a) Relative permittivity k, remanent polarization P_r, coercive field E_c and switching field E_s over ALD pulse ratio (dashed lines serve as a guide for the eye), b) to d) Large-signal P-E and small-signal C-E (stepwise field sweep via Keithley 4200-SCS) measurements from pure HfO$_2$ 3.4 cat% Sr and 7.9 cat% Sr, respectively. (reproduced from Schenk et al.[141])

Similar to what has been shown in the first publications on FE HfO$_2$[138, 132], the relative permittivity changes with doping. For pure, monoclinc HfO$_2$ a value around 20 is anticipated. Here, the value appears slightly higher ($\approx$ 24), which might either be due to contributions of an interfacial layer at the electrodes or because of a certain amount of orthorhombic phase being already present as evidenced by the small hysteresis in the P-E loop. Around the ferroelectric concentrations a clear butterfly-like shape of the small-signal permittivity vs. voltage is observed. Going to higher concentrations, the k-value is expected to drop again and to settle to a level anticipated for the higher symmetry cubic or tetragonal phases ($\approx$ 30). However, this is not the case in Fig. 4.9. The k-value drops again to around 23 – 24. This might by due to the low permittivity of SrO[128], which influences the effective k-values of the whole stack either via an "effective medium" consideration[247, 248, 249] or by segregation of

SrO at these high concentrations. An increasing amount of monoclinic phase with increasing Sr content is a tempting explanation for values similar to that of pure HfO_2 but it is not supported by the GIXRD data. The relative permittivity obtained by small-signal C-E measurements in the MIM stacks of this work is further discussed in section 5.6 via TEM and IS. Literature references for the permittivity values of the respective phases can also be found there.

Summary of the Dopant Influence

Both Si and Sr are able to induce ferroelectric behavior in HfO_2 thin films. The transition of the ferroelectric behavior is also accompanied by an evolution in the crystal structure. Whereas for the Si:HfO_2 films, a tetragonal phase could be proven, no clear distinction between a cubic and tetragonal phase at highest available concentration was possible for the Sr:HfO_2 films. On the one hand, this is due to the lower thickness of 10 nm instead of 30 nm. On the other hand, the used dopant concentrations were not high enough to suppress the FE properties again. Moreover, the ALD schemes of x:2 and x:3 for HfO_2:SrO supercycles applied for the two highest concentrations might not have been as effective as a x:1 scheme. From Fig. 4.10, it is evident that the concentration window for Sr doping is much wider than for Si (electrical data for 10 nm thick Sr and Si doped HfO_2 films). Besides the remanent polarization P_r, the saturation polarization P_{sat} is plotted to show the wide concentration range for which the field-induced phase transition is observed. As it has been pointed out in the above discussion of Fig. 4.7, a phase mixture is present in the Si:HfO_2 samples and the energetic differences between those phases seem small in the concentration range where the ferroelectric properties are observed. Ab-initio simulations of total energy E_{tot} (Fig 4.11) support this conclusion: Si is a much stronger stabilizer of the tetragonal phase than Sr. This strong stabilization of the tetragonal phase counteracts the anticipated benefit of doping, which is reducing the energy of the orthorhombic compared to the monoclinic bulk phase. If the tetragonal phase is stabilized by the influence of doping alone, there is no chance of utilizing surface energy contributions in favor of the orthorhombic phase. This is an illustrative example for the role of different dopants already outlined in Fig. 4.1 at the beginning of this chapter. If all other factors cause the Gibbs energies of the phases to become sufficiently close to each other, the impact of the dopant decides what phase becomes stable. Or the other way around, if a dopant has a really tremendous effect on total energy in the order of 50 meV or more, it might stabilize the orthorhombic phase without the need of surface energy and pave the way toward FE volume ceramics.

4.3 Impact of Smaller and Larger Dopants

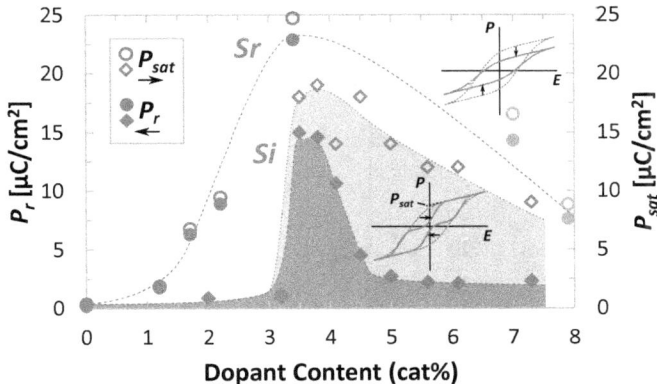

Fig. 4.10: Remanent and saturation polarization P_r and P_{sat} as a function of dopant content for 10 nm thick Sr:HfO$_2$ and Si:HfO$_2$: For Sr, both P_r and P_{sat} decrease simultaneously after a maximum is reached. In contrast, P_{sat} stays still at higher levels for Si even when the maximum in P_r is passed. This is due to a field-induced phase transition that occurs from a tetragonal paraelectric into the orthorhombic ferroelectric phase. The films with highest Sr contents were fabricated using different ALD supercycling schemes and are thus shown in lighter color.

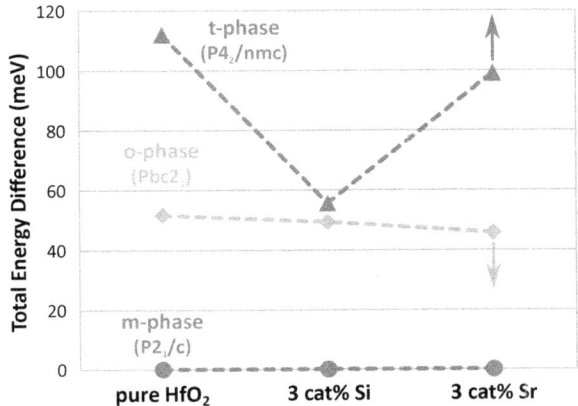

Fig. 4.11: Impact of Si and Sr doping on total energy from ab-initio simulations: While the relative energy reduction of the orthorhombic phase for both Si and Sr doping is favorable and opens a window for a stable FE phase, the strong energy reduction of the tetragonal phase counteract this effect for Si. The tetragonal phase is also favored for Sr doping, but this effect is by far less severe. The arrows indicate the desired situation of reducing the energy difference of the orthorhombic phase and increasing or keeping it constant for the tetragonal phase.[141]

4.4 Influence of Film Thickness and Annealing Conditions

For Si:HfO$_2$, the influence of film thickness and annealing conditions has been shown by Yurchuk et al.[250] and Lomenzo et al.[251]. For the binary mixture with ZrO$_2$, Park et al.[199] and Shimizu et al.[191] performed corresponding experiments. However, for the larger dopants, such a study has not existed so far and could prove whether there are general trends or not.

Also in the present work, trends in structural and electrical properties can be observed by increasing the thickness of the films as shown in Fig. 4.12. The same constant Gd concentration as for the anneal experiment was used and all three samples were annealed at 650 °C for 20 s. Instead of dropping as reported for other dopants[172], P_r still increased for the highest thickness used here. Interestingly, this is accompanied by a slight decrease in k and an increase in E_c. Although all films show comparable FE properties, the peaks around 18° and 25° gain intensity going from 10 nm to 27 nm thickness. Thus, the peaks might be attributed to the monoclinic rather than the orthorhombic phase. The latter, consequently, would have to have a texture that excludes these peaks in the GIXRD geometry. Moreover, there are hints at a decreasing fraction of the higher symmetry phases—i.e. a trend from a mixture of tetragonal/cubic and orthorhombic towards orthorhombic and monoclinic grains. The shoulders of the peaks around 30° and 35° become more prominent with increasing thickness a for 27 nm thickness. This is similar to the trend for lower Si concentrations in Fig. 4.6. Around 40° to 45°, 55° and 65° to 80°, multiple orthorhombic or monoclinic peaks appear for increasing thickness.

Fig. 4.13 shows the diffraction data and polarization hystereses for TiN-Gd:HfO$_2$-TiN stacks subjected to N$_2$ anneals ranging from 450 °C, 10 min up to 800 °C, 20 s. All P-E curves were measured after a pre-conditioning sequence of 10^4 cycles at 10 kHz. They show marked FE properties for all annealing conditions used. Together with the anneal at 1000 °C, 1 s reported by S. Mueller[139], Gd:HfO$_2$ possesses the widest anneal temperature range for FE properties shown for HfO$_2$/ZrO$_2$ based films so far. 1000 °C, 1 s would be a typical thermal budget for a film that has to endure a subsequent activation anneal for the source and drain implants, whereas 450 °C, 10 min represents a thermal budget suitable for a crystallization within the so-called back-end of line process chain, which is typical e.g. for the capacitor fabrication in state-of-the-art DRAM routes or so-called gate-last process schemes[252, 253]. The high temperature anneals were shown for many other dopants as well[140]. Up to now, a temperature of 450 °C was shown to induce ferroelectricity only in the binary mixture HfO$_2$-ZrO$_2$, which, however, did not allow for anneals above 650 °C.[191]

4.4 Influence of Film Thickness and Annealing Conditions

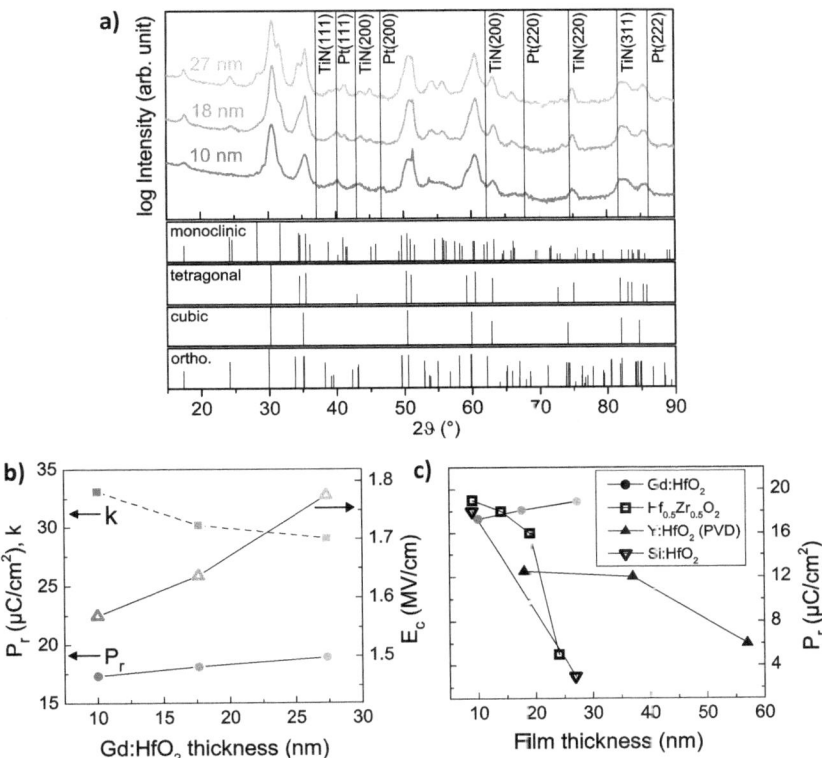

Fig. 4.12: a) GIXRD patterns for different Gd:HfO$_2$ film thicknesses and HfO$_2$ reference patterns; b) relative permittivity k, remanent polarization P_r, and coercive field E_c as a function of Gd:HfO$_2$ thickness; c) evolution of P_r with film thickness compared to other dopants. (reproduced from Hoffmann et al.[172])

However, there are some differences between the four anneals shown in Fig. 4.13[254]:

- The two anneals with highest thermal budget—800 °C for 20 s and 650 °C for 10 min—seem quite comparable in the P-E curves as well as in the diffractograms (compare the discussion of the diffractograms given above for the thickness variation). The same applies to the two anneals at lower thermal budget—650 °C for 20 s and 450 °C for 10 min.
- The high thermal budget anneals result in an E_c of 1.8...1.9 MV/cm. For the lower thermal budgets, it was 0.3 MV/cm lower.
- The 450 °C anneal results in the lowest P_r of 15 µC/cm^2 instead of the 18 µC/cm^2 achieved for the other anneals. As evident from GIXRD, this is not due to a higher amorphous fraction, which would give rise to an increased underground intensity, broader peaks of lower maximum intensity or even a halo (broad hump around the region of the main peaks, stemming from nanocrystalline regions formed at the beginning of

crystallization). However, since the P_r difference is rather small, a high fraction of orthorhombic FE phase has to be present in all four samples.
- The slope in the saturated/non-switching part of the hysteresis is higher for the anneals with lower thermal budget, which hints at a higher k-value. This is in accordance to the diffraction results, which hint at increased tetragonal/cubic fractions compared to the higher thermal budget anneals, which exhibit more monoclinic phase fractions (in addition to the expected orthorhombic grains).

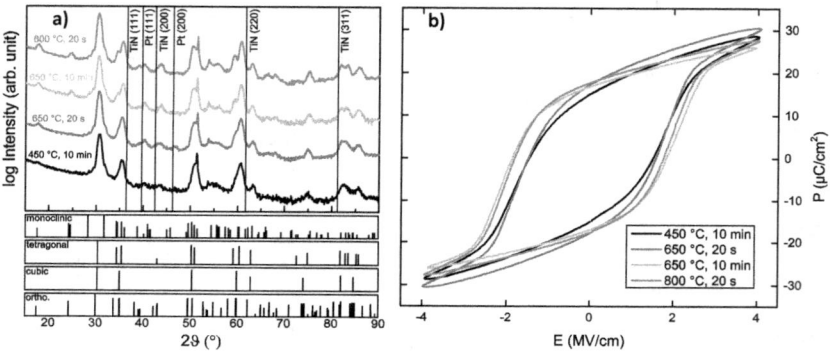

Fig. 4.13: a) Grazing incidence X-ray diffractograms for Gd:HfO$_2$ samples and reference patterns are given for the paraelectric monoclinic, tetragonal and cubic phases as well as for the ferroelectric orthorhombic phase. b) Comparison of the polarization hystereses resulting from different annealing conditions after 10^4 field cycles.[254]

The diffractograms can be discussed similar to what has been done for the thickness variation in Fig. 4.12 above. The effect of the anneals is less prominent though. However, they provide indications to decide whether the high symmetry phase induced by Gd doping is tetragonal or cubic. The peak around 35° shows a shoulder for the anneal at 800 °C for 20 s, which vanishes going towards lower thermal budget. This is similar to the trend for higher Si concentrations in Fig. 4.6. However, for higher Si concentrations a second peak appeared where the monoclinic peak shoulder was found before. This is not the case here. But given the limited effect of the anneals compared to the concentration variation in Fig. 4.6, this observation is not enough to clearly rule out the tetragonal and proof the presence of the cubic phase. The distorted and broad peak around 50° becomes narrower and a two- instead of a three-peak structure becomes present at around 60°. Altogether, these features are more supportive for the claim of a cubic rather than a tetragonal phase. Permittivity values provided in literature do not allow for a clear distinction either. Reported k-values for the cubic phase range from 29 to 39 and are included in the span of 28 to 70 given for the tetragonal phase (see Fig. 5.19 in section 5.6 for an overview corresponding references).

The high tolerance for annealing conditions and the absence of a marked drop in P_r as seen for other dopants suggests that Gd is very effective in stabilizing the FE phase. For Hf$_{1-x}$Zr$_x$O$_2$ with high amounts of zirconia, the crystallization temperature is below 300 °C and reduces

4.5 Electrode Effects

further for thicker films. A partial or even complete crystallization into a non-polar phase during deposition has to be considered when interpreting corresponding data.

4.5 Electrode Effects

As already mentioned in section 2.4, electrode interfaces (and interfaces in general) play an important role with respect to polarization discontinuity.[31, 29, 73, 74, 48] However, those effects are not the central topic of this section. The electrode effects discussed here can be divided into structural and chemical effects.

Structural effects can be understood from Fig. 4.14. The dark field image in Fig. 4.14 a) exemplarily shows that only about two to three neighboring grains are similarly oriented (bright appearance: the grains fulfill the conditions for constructive interference with the same family of lattice plane); hafnia grains span the whole film thickness, but are limited to about two to three TiN grains in lateral direction with grain boundaries coinciding. Moreover, the bright field image in Fig. 4.14 b) shows that there is an orientation relationship between the HfO_2 and the TiN lattice for some of the grains.

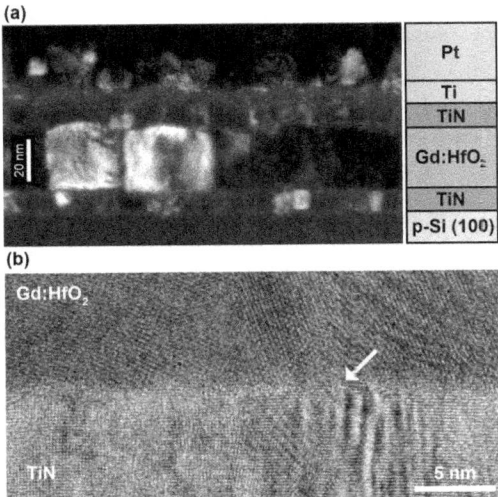

Fig. 4.14: a) Dark field TEM shows $Gd:HfO_2$ grain sizes of about 27 nm. b) Bright field TEM of the BE interface. The arrow indicates a $Gd:HfO_2$ grain boundary which coincides with a TiN boundary.[172]

A **chemical effect** was derived from a comparison of TiN and TaN for 10 nm thick $Gd:HfO_2$ films of same dopant concentration. Fig. 4.15 shows the diffractograms and the electrical characteristics P_r and k for different electrode combinations and with varied anneals. Similar to what was argued in section 4.4, an evolution in the structural data is found: Higher thermal budgets cause the diffractograms to change from more monoclinic and orthorhombic to orthorhombic and tetragonal/cubic. In general, it seems, the TaN/TaN (top/bottom electrode) samples possess a lower monoclinic fraction compared to the TiN/TiN samples in Fig. 4.13.

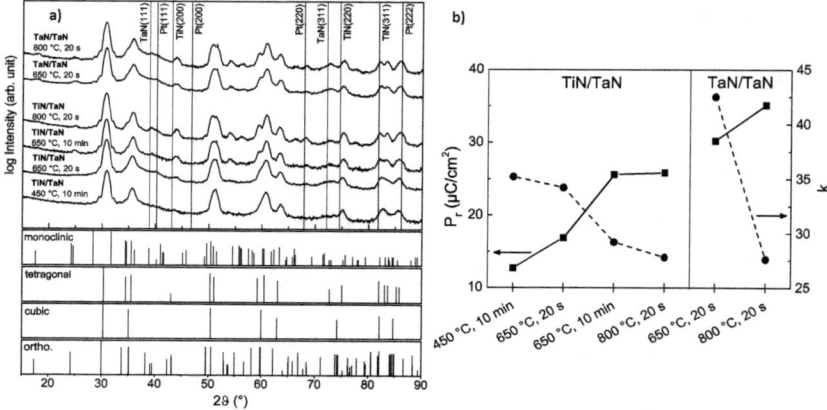

Fig. 4.15: Example of Gd:HfO2 (3.4 cat%): a) GIXRD measurement and reference patterns and b) relative permittivity k and remanent polarization P_r for different TE/BE configurations and annealing conditions.[172]

The electrical data shows a strong dependence on the annealing conditions when a TaN electrode is involved. The remanent polarization increases from $13\,\mu C/cm^2$ to $35\,\mu C/cm^2$ between TiN/TaN at 450 °C for 10 min and TaN/TaN at 800 °C for 20 s. The latter means a doubling of P_r compared to the standard case with TiN/TiN and is the second largest value reported up to now—$40\,\mu C/cm^2$ was achieved with La:HfO$_2$ with TiN electrodes[140]. Similar to previous reports[254, 191], k decreases with increasing thermal budget. The k-value of 43 achieved for 650 °C for 20 s hints at significant fractions of the higher symmetry phases. It strongly drops going to the 800 °C, while P_r still increases by $5\,\mu C/cm^2$. This might point to more pronounced interface growth, which reduces the effective k-value of the film. This temperature dependence of k seems stronger than for the TiN/TiN or TaN/TaN samples. Hoffmann et al.[172] also showed an earlier dielectric breakdown for the samples with TaN electrodes. Leakage currents for the samples were reported to be comparable. A higher P_r itself gives rise to higher displacement currents and depolarization fields and causes a higher stress during cyclic switching of the capacitor stacks. However, it was speculated that the oxidation affinity of TaN could be higher than that of TiN. Time-of-flight secondary ion mass spectrometry (ToF-SIMS)[154] was applied to check this hypothesis. Fig. 4.16 shows the results giving hints to support this hypothesis:

Firstly, it is important to mention that there are false (order of magnitude higher than caused by the actual presence of these species) TaO_2^- and TaN^- signals in the Gd:HfO$_2$ film with TiN electrodes. Similarly false is the HfO$_2$ signal at the surface of the surface of the TaN/TaN sample. However, there seems to be some TaO_2^- intensity left throughout the whole HfO$_2$

4.5 Electrode Effects

layer with TaN TE and BE when subtracting the TaO_2^- background of Fig. 4.16 a) from that of Fig. 4.16 c). The edges in the traces of TaN^- and TaO_2^- appear less steep than in the traces of TiN^- and $^{50}TiO^-$. Additionally, there is a strong oxidation (TaO_2^- signal) visible at the air exposed surface of the TaN top electrode.

Nonetheless, artifacts such as the steps in the TiN^- signal in the middle of the TiN electrodes and interfering masses of isotopes (e.g. HfO_2^- and TaO_2^-; TiN^- and $^{50}TiO^-$, TaN^- and HfO^-) occur. Thus, the ToF-SIMS can only serve as a first indication.[172] Ab-initio calculations were able to provide further support. Oxygen vacancies (charges compensated within the unit cell) were incorporated at different lattice positions and an average effect on total energy was calculated. As Fig. 4.17 shows, the monoclinic phase can be suppressed by oxygen vacancies. The ≈ 30 meV per formular unit (f.u.) for 12 at% oxygen vacancies equal half the difference between monoclinic and orthorhombic phase in ground state as mentioned in section 4.1 (Fig. 4.1). This effect is in in accordance with the experimental results discussed above. Affecting the oxygen vacancy concentration in the ferroelectric and with it the phase stability adds a chemical effect of the electrodes to the structural effects presented at the beginning of this section.

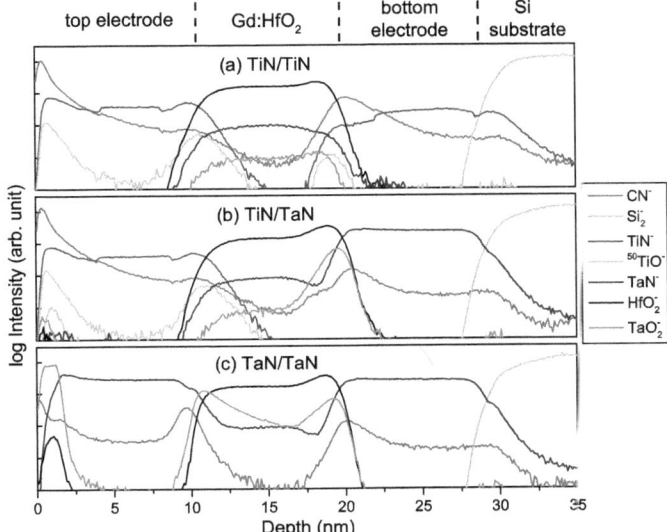

Fig. 4.16: ToF-SIMS results for a) TiN/TiN, b) TiN/TaN, and c) TaN/TaN samples annealed at 650 °C for 20 s. Bars on the y-axis indicate a factor of 10 in intensity ratio.[172] The SIMS measurements were performed by Maximilian Drescher at Fraunhofer IPMS, Dresden.

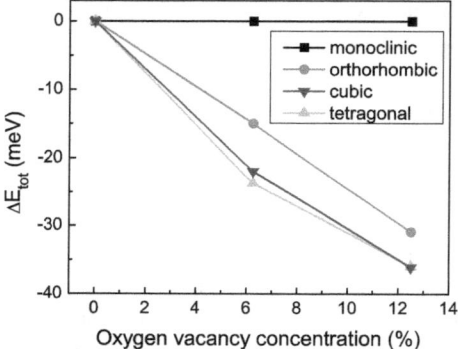

Fig. 4.17: Average change in the total energy difference ΔE_{tot} between the various phases of HfO$_2$ compared to the monoclinic phase as a function of the oxygen vacancy concentration.[172]

4.6 Texture of the Ferroelectric Thin Films

Texture of polycrystalline samples defines the effective values of polarization and coercive fields. Many publications included GIXRD (with line focus) as part of their standard structural characterization. Thin films with thicknesses in the order of a few ten nanometers hamper the use of a classical Bragg-Brentano geometry as described in section 3.2. This also affects the assessment of texture. A preferential texture with the polar c-axis perfectly out-of-plane, i.e. perpendicular to the electrode interfaces would maximize the sum of the projected polarization contributions and thus, P_r as the macroscopically measured film property.

In a first step, it was proven by GIXRD that there is no significant difference in preferential orientation when sample is rotated around the normal of the film plane (angle Φ) as can be seen from Fig. 4.18. All 2Θ-scans at different Φ look nearly the same. This means the film exhibits what is called a fiber texture[255, 156].

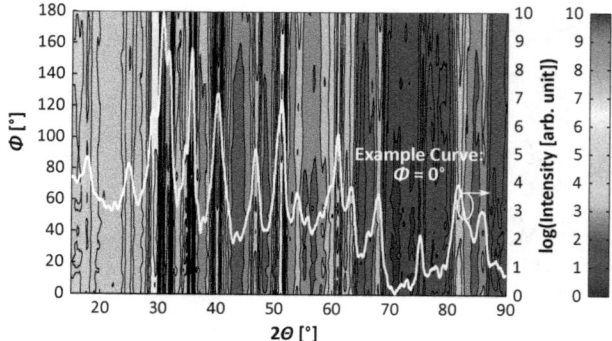

Fig. 4.18: GIXRD scans for different rotations around the plane normal (angle Φ) in fine steps of 5° for $\Phi = 0°...60°$ and coarse steps of $\Phi = 15°$ for 60°... 80°. No significant in-plane texture is observed. This sample (Si:HfO$_2$, ALD cycle ratio of 26:1, thickness of 30 nm) exhibits a so-called fiber texture.

4.6 Texture of the Ferroelectric Thin Films

The proof of a fiber texture is a prerequisite that eases the subsequent analysis by reducing the number of required time consuming scans that have to be performed in a geometry less suitable for thin films. These coupled 2Θ-Θ-scans were performed next. The Bragg-Brentano geometry makes it easier to identify the orientation of the planes contributing to the observed peaks. As the stray vector is fixed during the whole scan, all crystal plane normals are parallel to the stray vector and one another. Changing the angle Ψ between stray vector and film normal (angle of the Eulerian cradle) and repeating the 2Θ-Θ-scan allows identifying the orientation of the polar [001] crystal direction of the FE phase. Besides the way of scanning, the setup was not changed to maximize the intensity for the only $\approx$ 30 nm thick film.

Fig. 4.19 shows the respective patterns of these 2Θ-Θ-scans and the corresponding results of their simultaneous Rietveld refinement using MAUD[256, 257] (version 2.7). This methods allows the identification of an orientation distribution function using a spherical harmonic model. A weak and mixed texture of only up to $\approx$3 multiples of a random distribution (MRD) was found for all phases (Fig 4.19). This means that the polar [001] axis is nearly equally distributed over all possible orientations in space. Assuming a perfectly equal distribution, the macroscopically effective remanent polarization can be estimated as

$$\frac{P_r}{P_s} = \frac{2}{\pi} \cdot \int_0^{\pi/2} \cos(\alpha)\,d\alpha = \frac{2}{\pi} \cdot [\sin(\alpha)]_0^{\pi/2} = \frac{2}{\pi} \approx 0.64 \tag{4.1}$$

with α being the angle between film normal and orthorhombic [001] direction. As a consequence, just a $P_r = 33 - 34\,\mu C/cm^2$ instead of $P_s = 52 - 54\,\mu C/cm^2$ as calculated from ab initio simulations is expected if all grains would exhibit the FE phase. Considering the refined orthorhombic phase fraction of $0.4/(0.4 + 0.25) \approx 60\,mol\%$, the expected P_r would drop to around $20\,\mu C/cm^2$. Compared to the value of $12\,\mu C/cm^2$ shown in Fig. 4.6, this is still too large. However, this pristine hysteresis is significantly rounded and the saturation polarization represents a better reference value not afflicted by polarization relaxation (dead layer effects, remaining internal bias fields,...). Extrapolating from the saturated linear branches toward zero field gives a value for the saturation polarization P_{sat} of $\approx 18\,\mu C/cm^2$, which is in good agreement with the XRD based estimate.

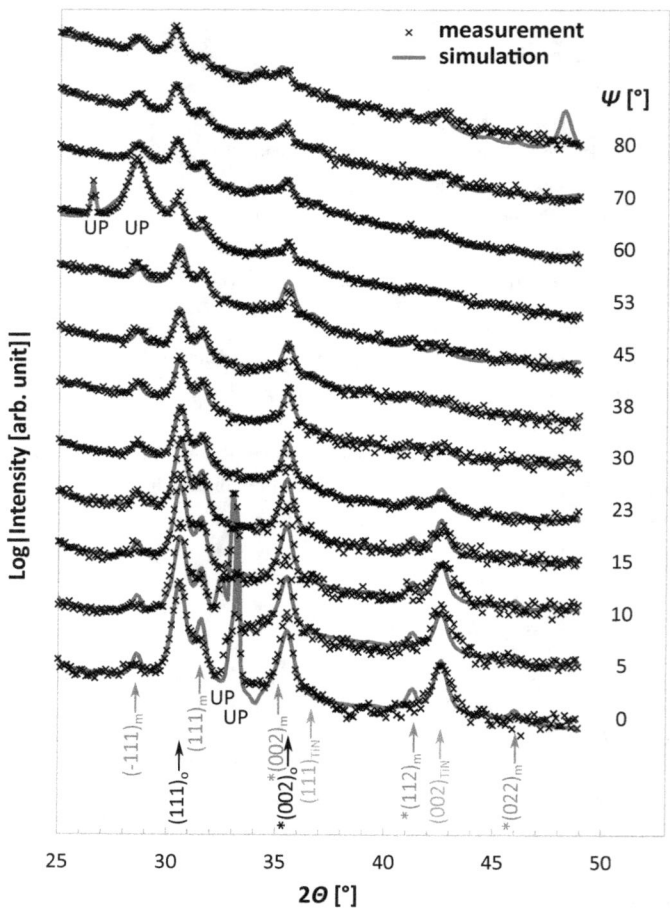

Fig. 4.19: Rietveld refinement of 2Θ-Θ-scans for different Ψ (Si:HfO$_2$, ALD cycle ratio of 26:1, thickness of 30 nm): Phase fractions of 25 mol% monoclinic, 40 mol% orthorhombic HfO$_2$ and 35 mol% TiN were obtained. The orthorhombic phase was approximated with a tetragonal structure to reduce the amount of fit parameters in the refinement. A polynomial background fit and a spherical harmonic texture model was used. Moreover, underground peaks (UP) were introduced to account for substrate-related and other artifacts such as double diffractions or fluorescence phenomena. For the sake of convenience, the location of main peaks of all three phases are shown in the diagram in a simplified manner. The * symbol indicates the presence of multiple non-equivalent families of lattice plane that cause reflexes at different locations around the position marked by the respective arrow. The Rietveld refinement was performed by Christopher M. Fancher (CNMS, Oak Ridge National Laboratories).

A disadvantage of the texture model used is that the MRD values are not forced to be greater than zero as evidenced by the black areas especially in the (002) pole figure of the monoclinic phase in Fig 4.19. The orthorhombic phase was approximated with the tetragonal phase. This means, [002], [110] and [−110] directions of the tetragonal cell correspond to the main lattice vectors a, b, and c of the orthorhombic cell. However, oxygen positions do hardly contribute to the XRD signal and thus, the equivalence is not well-defined.

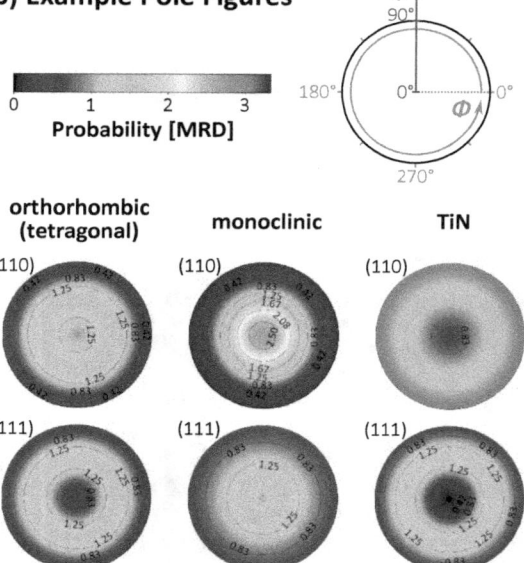

Fig. 4.20: Texture assessment: a) Sketch of the angles in the used Bragg-Brentano geometry; b) Example pole figures reconstructed from the texture spherical harmonic model show only a weak mixed texture of up to 3 MRD (multiples of random distribution). Within the uncertainties of the refinement, the results do not justify to conclude that a significant preferential orientation is present in the HfO$_2$ or TiN and the orientation is rather random.

The (111) peak of the FE phase moves significantly between the patterns obtain from different Ψ (see Fig. 4.19). The strain between in- and out-of-plane direction is around 1 % as can be seen from the slope and y-axis intercept in Fig. 4.21. From these peak shifts in this plot of lattice plane spacing $d\,(\Psi)$ vs. $\sin^2(\Psi)$, the residual stress in the thin film could be calculated of via the so-called $\sin^2(\Psi)$ method[258], if the elastic properties of the phase were known. However, no such values have been published for the FE phase yet. Interestingly, no such peak shift was found for the monoclinic peaks and the (002) peak of the orthorhombic phase. As argued by [1], the formation of the orthorhombic phase during annealing is supported by a clamping effect of the TiN electrodes. This clamping prevents the shearing of the metastable tetragonal phase, which is necessary to form the monoclinic phase. The found residual stress along the space diagonal could be a residue of this process Up to now, only hydrostatic, uniaxial, and biaxial stress conditions along main crystallographic axes have been considered.[61, 160] The results found here suggest follow-up studies to prove if other thin film samples behave similar and additional simulation efforts in the direction of more complex stress conditions.

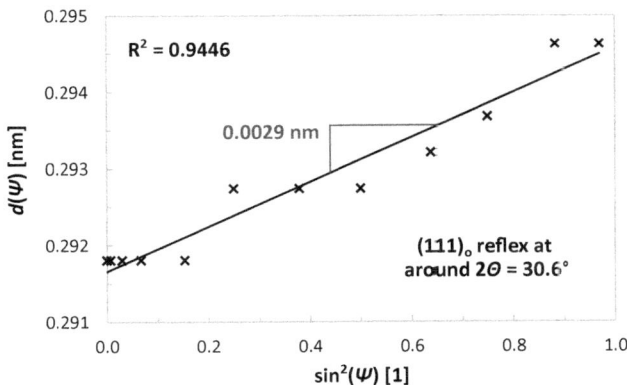

Fig. 4.21: Characteristic plot of the $\sin^2(\Psi)$ method[258]: Only the (111) peak of the FE phase exhibits a significant shift between in- and out-of-plane orientation. The difference in lattice plane spacing is found to be around 1 %. If elastic properties were known, the residual stress could be calculated from the plot.

4.7 HfO$_2$—An Incipient Ferroelectric?

To round up the section of stabilizing the FE phase in HfO$_2$, the claim of HfO$_2$ being an "incipient" ferroelectric should be discussed. This term has been used by S. Mueller et al.[138] in one of the first publications trying to integrate the new observations of HfO$_2$ into the existing picture of phenomena for conventional ferroelectrics. Going to ultra-low temperatures, incipient ferroelectrics show a suppressed transition from a paraelectric to a ferroelectric state. The transition is only suppressed by quantum fluctuations.[259, 260, 261, 262] The term "incipient" refers to the fact that there is only an "onset" of ferroelectricity but the materials does not completely overcome the barrier to a stable ferroelectric state. In order to observe such behavior, the energy difference between the paraelectric and ferroelectric state, which reduces during cooling, has to be low enough. Fig. 4.22 shows the different behavior of χ and $1/\chi$ compared to what has been discussed in section 2.2 (Fig. 2.4 and Fig. 2.5). The **transition temperature T_0 is close to 0 K or slightly below**, but quantum fluctuations prevent the expected case of $\chi \to \infty$ for $T_0 \leftarrow T$.

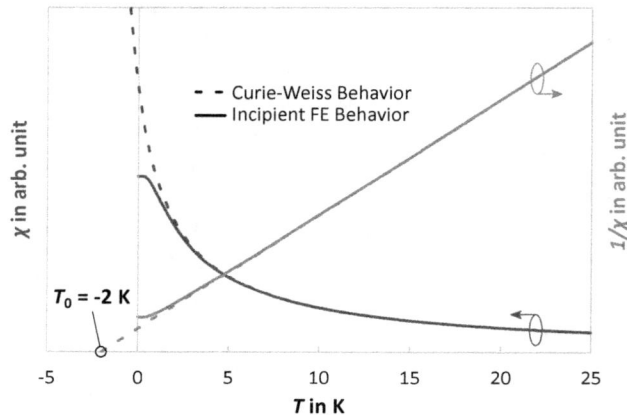

Fig. 4.22: Sketch of incipient ferroelectricity: Temperature dependence of dielectric behavior anticipated from Curie-Weiss law (Eq. 2.12) and deviation from that case for incipient ferroelectricity (modeled as described by Barrett et al.[263])

At this point, any further discussion becomes philosophic: What should be counted as "close to or slightly below 0 K"? What other boundary conditions are legitimate to attribute the behavior to a material property instead of e.g. a film property or, more general, a sample property?

Bulk HfO_2 is not expected to show such behavior because the low temperature (lower symmetry) monoclinic phase is already stable at room temperature and will gain further stability advantage upon cooling. A transition from a high temperature paraelectric to a low-temperature ferroelectric phase is needed as described in section 2.2.

Playing with the thickness of pure HfO_2 films, Polakowski et al.[198] showed FE properties arising for thicknesses below 10 nm. Going to 8 nm and 6 nm, the P_r still increased. If the material is already FE at room temperature, it cannot undergo a transition from para- to ferroelectric upon cooling. Such a sample will, thus, also not show the behavior typical for an incipient ferroelectric.

Using even thinner HfO_2 films to promote the surface energy contributions or a sufficiently high concentration of a suitable dopant, a high temperature phase (cubic or tetragonal) is stabilized at room temperature. In this case, the observation of incipient ferroelectricity is possible. If the film is too thin or if the dopant concentration is too high, the energy difference to the orthorhombic FE phase becomes too large to see any onsetting stabilization of a ferroelectric phase. These are just two knobs that can be utilized to push the transition temperature into a suitable range for what is shown in Fig. 4.22. These types of manipulation are also known for conventional ferroelectrics. $SrTiO_3$ is a typical example of materials referred to as incipient FE. Also for this material a FE phase can be induced by defects (even changing isotopes) or certain stress conditions.[264, 265, 266, 267]

Summing up, pure bulk HfO_2 cannot be called an incipient ferroelectric in a narrower sense of the word. The term "incipient" depends solely on a low energy barrier between a paraelectric room-temperature phase and a potential lower-temperature ferroelectric phase. If other "engineering" knobs, such as surface energy, doping, stress, ... have to be applied to lower this barrier to make "incipient ferroelectricity" observable, the use of a broader sense of the term "incipient ferroelectricity" as a sample instead of a material property has to be debated.

4.8 Summary

Right at the beginning of this chapter, it has been argued that ferroelectricity in hafnia is a quite universal behavior since it was established in films fabricated by a wealth of different deposition techniques, incorporating many different dopants, pure HfO_2 or using a binary mixture with ZrO_2 and sandwiched between various electrode materials.

The main restriction to this "universality" is the limitation of the observations to thin films only. **Surface energy** seems to be a main knob to create a window between the monoclinic bulk phase and the high symmetry tetragonal or cubic phase in thin films. A clear trend for more monoclinic toward more tetragonal/cubic has been presented in section 4.4. For ceramics with typical grain sizes of $> 1\,\mu m$, surface energy is negligible and cannot be utilized anymore.

Other stabilizers are needed. An increasing thermal budget was found to have a similar effect as an increase of the film thickness . Dopants can be used to create additional disorder, favoring the higher symmetry phases. An empirical model is sketched in Fig. 4.23 summarizing the aforementioned trends for the **influence of process parameters**. Film thickness and thermal budget act via the lever of surface energy, but during the anneal, the grain growth and a transformation from a high symmetry phase into the orthorhombic FE phase have to be considered.

Dopants of strongly different ionic radius compared to Hf have been studied. As representatives of the bigger dopants, Gd^{3+} as lanthanoid and Sr^{2-} as an alkaline earth metal have been used and proven to be more suitable than the originally used Si^{4+} for opening a window for ferroelectric behavior (Fig.4.10). Higher concentration and film thickness windows have been demonstrated. In contrast to $Si:HfO_2$, higher dopant concentrations lead to a more cubic-like phase—either a cubic one or tetragonal one with low lattice constant differences.

Silicon, has clearly been shown to be a strong stabilizer of the tetragonal phase (Fig 4.11). Since this phase is a parent phase of the orthorhombic phase, the barrier is expected to be low and a **field-induced phase transition** is observed, explaining the observation of AFE-like behavior similar to that in Al:HfO$_2$ and Hf$_{1-x}$Zr$_x$O$_2$. Due to the low concentration window for the FE phase, a very sensitive multiphase coexistence has been shown in Si:HfO$_2$ films. Besides the dopant ions themselves, the automatically created oxygen vacancies impact the phase stability and, as defects, favor higher symmetry phases.

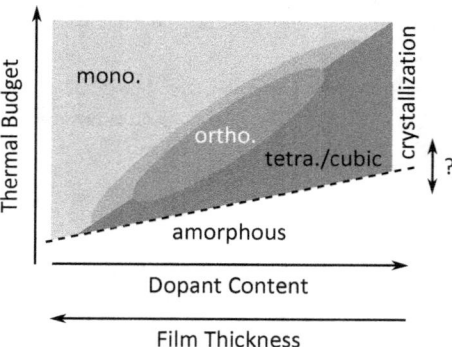

Fig. 4.23: Qualitative model of the phase transitions of HfO$_2$ influenced by the most important process parameters.[172]

A **structural and a chemical effect of the electrodes** have been demonstrated for Gd:HfO$_2$ sandwiched by TiN or TaN. The lateral grain size forming during annealing is influenced by the already present grain structure of the electrodes. Additionally the grain structure seems to impact the preferential orientation of the later on crystallized FE film. These are the structural impacts. A chemical component is given by the capability of pulling oxygen out of the FE layer, which affects the phase stability and can—to a certain degree—promote the FE properties.[6]

A fiber **texture** (insensitivity of the measured pattern to rotations around the film plane normal) has been shown in the FE capacitor structures. Additionally, not significant out-of-plane texture has been found by XRD measurements. This nearly random distribution together with the refined orthorhombic phase fraction of 60 mol% accounts for the observation of $P_{sat} \approx 20\,\mu\text{C/cm}^2$ instead of the theoretical values of $\approx 52\,\mu\text{C/cm}^2$ in the studied Si:HfO$_2$ sample.

Finally, it was discussed that pure, bulk HfO$_2$ cannot be called an incipient ferroelectric in a narrower sense of the word. Similar to what has been done for SrTiO$_3$, defects, stress, surface energy,... can be used to achieve a behavior similar to that of an incipient ferroelectric in certain samples.

---

[6]Of course, ferroelectric instability at the boundaries of the FE film, as it has been discussed already in the 1950s[29, 73, 74], should not be ignored. A sufficient screening of the polarization charge is another important role of the electrodes.

5 Electric Field Cycling Behavior of Ferroelectric Capacitors

The term "electric field cycling behavior" is used to describe the reaction of the ferroelectric material to alternating pulses of the electric field to (partially) switch the material. It condenses three different effects as illustrated in Fig. 5.1 below[48, 187]:

1) "Wake-up": The pinched hysteresis of a pristine (= as-fabricated) sample opens during cyclic switching.[71, 36] This usually is also accompanied by a more or less prominent increase in P_r and is sometimes also called "deaging"[38, 42, 43] for conventional ferroelectrics.
2) Polarization fatigue: During continued field cycling, P_r starts decreasing, which is related to a reducing amount of domains that are still able to participate in the switching process.[58, 59]
3) "Split-up" effect: Subjecting a sample with an open (depinched) hysteresis to a sequence of switching pulses of non-saturating amplitude (subcycling), the transient current peaks split-up around the subcycling amplitude, which is accompanied by a constriction of the hysteresis. This effect can be reversed by again applying a field cycling sequence of saturating amplitude.[48]

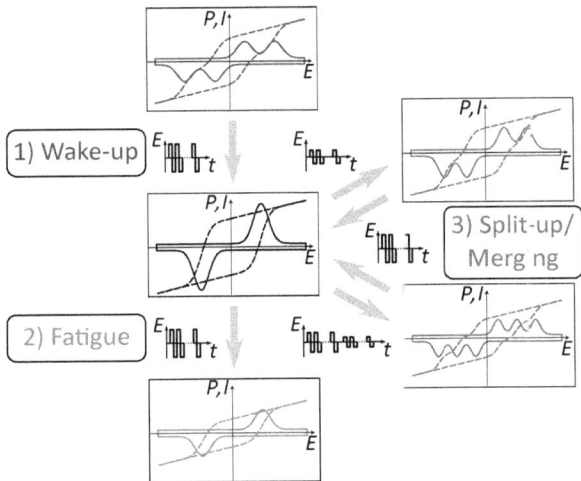

Fig. 5.1: Schematic of the experimental observations for the three phenomena condensed in the term "electric field cycling behavior".[187]

All three phenomena are described for the example of Sr:HfO$_2$ (3.4 cat% doping) in a similar way in their respective subchapters: First, the manifestation in P-E and I-E curves (compare Fig. 2.6 in section 2.4) and the main findings of harmonic analysis are explained. Since these P-E and I-E results can only serve as a first indicator, the corresponding Preisach or switching densities obtained from first-order reversal curve (FORC) measurements are discussed. The FORC approach was chosen because different switching density plots are expected for different scenarios discussed in section 2.4 as it is contrasted in Fig. 5.2. Later on, the obtained activation energies are discussed and SPM, TEM and IS efforts are presented trying to resolve some microscopic origins of the electric field cycling behavior.

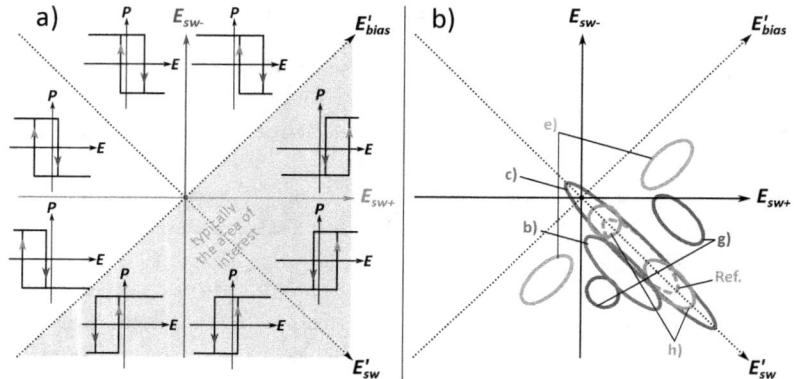

Fig. 5.2: a) Preisach plane and the switching fields of the respective bistable units in different sections of the plane. b) Sketch of the contour-plots to be observed for different scenarios from Fig. 2.4. The ellipses show the location of a maximum in the switching density. ($E_{sw+/-}$ denote the field necessary to switch to positive/negative polarization state. E'_{sw} and E'_{bias} are transforms of alternative coordinates for the Preisach plane that represent the mean absolute value of the switching field and the internal bias field, respectively).[48]

5.1 Wake-up Effect

As-fabricated, i.e. uncycled samples often exhibit two or more distinct peaks in the transient current characteristic, which results in a pinched polarization hysteresis. During an increasing number of switching cycles at saturating field, these maxima move towards each other ending up merged into one single peak as Fig. 5.3 shows for a temperature of 298 K and a frequency of 1 kHz.

The experiment was repeated using different frequencies and different temperatures between 136 K and 298 K. A modified approach of harmonic analysis as described in section 3.7 was used to determine activation energies. Fig. 5.4 exemplarily shows the evolution of the harmonic's amplitudes and phase angles for $f = 10\,\text{kHz}$ and $T = 171\,\text{K}$ and the corresponding data points in the Arrhenius plot obtained from the extracted phase jump at this specific temperature. Activation energies between 84 meV and 126 meV were extracted. An overview of all activation energies for wake-up, fatigue and split-up is given in section 5.4, which also discusses their frequency dependence and the dependence on the different harmonics.

5.1 Wake-up Effect

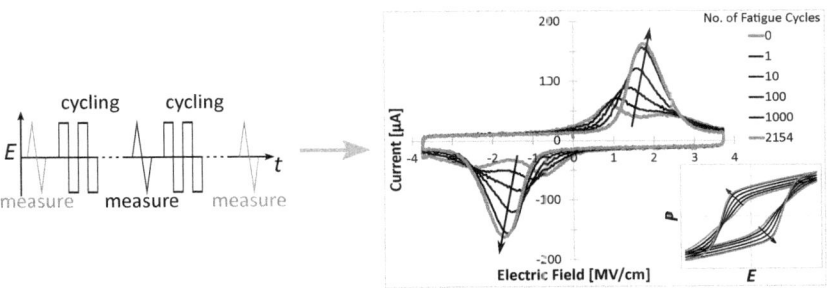

Fig. 5.3: Wake-up effect: The constricted hysteresis of a pristine sample opens after field cycling and two distinct current peaks merge into one single peak. The corresponding P-E hystereses are shown as inset with the E- and P-axes scaled from $-4\,\text{MV/cm}$ to $4\,\text{MV/cm}$ and from $-30\,\mu\text{C/cm}^2$ to $30\,\mu\text{C/cm}^2$, respectively. The cycling sequence is sketched next to the graph.[48]

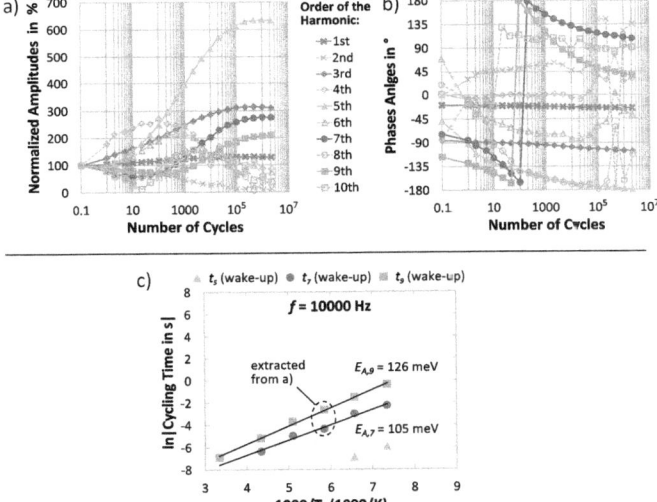

Fig. 5.4: Extraction of activation energies via harmonic analysis: a) Example of the evolution of amplitude (normalized to the respective initial value) and b) phase against the number of fatigue cycles for $f = 10\,\text{kHz}$ and $T = 171\,\text{K}$. Jumps in the phase of the seventh and ninth harmonics occur subsequently and are accompanied by a minimum in the amplitude. c) Arrhenius plot of the cycling time t_i at which the phase jumps of the respective i-th harmonic was observed. Data points were extracted from a) as shown and similarly for the other temperatures between $136\,\text{K}$ and $298\,\text{K}$.[48]

In order to distinguish, if cases g) or h) from Fig. 5.2 account for the initial double peak structure, FORC measurements were performed after different numbers of switching cycles. Fig. 5.5 shows the switching density plots and corresponding P-E hystereses for a sample in a) pristine state, b) after 100 (somewhere in the middle of the wake-up process) and c) after 10^4 cycles (toward the end of wake-up) at $4\,\text{MV/cm}$ and $10\,\text{kHz}$ at $298\,\text{K}$. A comparison to Fig. 5.2

clearly shows that internal bias fields account for the initially constricted hysteresis. This proves that the origin of the pinching is of similar nature as in conventional ferroelectrics[41, 268] and that similar defect types should be considered in further studies: grain boundary, domain wall, or volume effects as summarized in section 2.4.

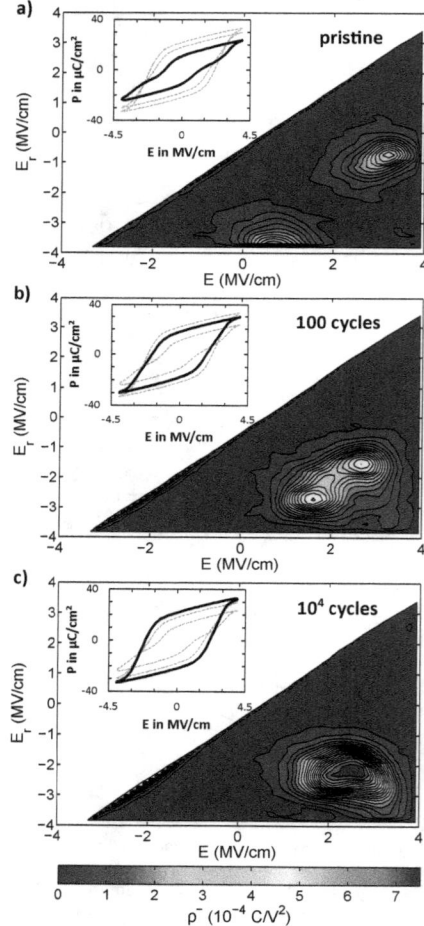

Fig. 5.5: Wake-up for 4 MV/cm amplitude: Experimental switching density determined for a) a pristine capacitor, b) after 100, and c) after 10^4 rectangular switching cycles. The insets show the respective polarization hystereses (extracted from the last sweep of the corresponding FORC measurement) for comparison. The solid line of each inset is the hysteresis for the accompanying switching density plot, whereas the dashed lines are the respective hystereses of the two other cycling stages for comparison.[187]

Targeting the application of ferroelectric memories, another aspect has to be discussed. As explained in section 2.5, compromise between data retention on the one hand and longest

5.1 Wake-up Effect

possible endurance or low-power operation on the other hand might be necessary. Thus, a subloop operation of a ferroelectric memory based on HfO_2 might be taken into account. Fig. 5.6 a) shows the endurance behavior (switchable polarization $2P_r$ vs. no. of switching cycles) for different field amplitudes. For 4.5 MV/cm, an initial and maximum $2P_r$ of 34 µC/cm^2 and 47 µC/cm^2, respectively, can be achieved. However, subjected to such high fields, the capacitor only withstands 100 cycles. Reducing the amplitude to 3.0 MV/cm, which is still well above the coercive field (compare Fig. 5.3 or Fig. 5.5), the sample is able to withstand 10^6 cycles. However, the initial and maximum $2P_r$ tremendously drop to only 5 µC/cm^2 and 15 µC/cm^2. Further decreasing the field amplitude to 2.25 M/cm, which is in the order of the coercive field, the $2P_r$ values are below 3 µC/cm^2 throughout the whole course of cycling operation. These strong reductions seem rather surprising at the first glance, but Fig. 5.6 b) illustrates the origin: The internal bias fields shift the maxima in the switching density outside the accessible ranges of the respective field amplitudes. Part of the domains (lower left maximum) can only be switched to positive polarization, but not back to the negative and the other way around for the oppositely biased domains (upper right maximum).[187] This problem has to be addressed to maximize the amount of switchable polarization at reduced voltages.

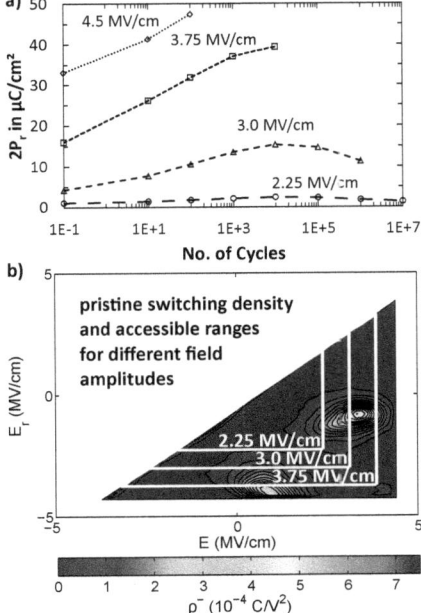

Fig. 5.6: Problem of internal bias fields for low-power-operation of ferroelectric memories: a) Evolution of $2P_r$ for different field amplitudes as a function of the number of cycles and b) switching density of the initial sample obtained with a FORC amplitude of 4.5 MV/cm. The triangular white frames represent the accessible range of domains for the respective lower field amplitudes. Internal bias fields decrease the amount of domains switchable in both polarization directions.[187]

It is important to note that this study is more of a fundamental than of a close-to-application character. At these rather low frequencies of 10 kHz, the endurance is also comparably low and limited by dielectric breakdown. This problem relaxes for higher frequencies. As a very rough estimate (if solely the accumulated time of the applied amplitude counts): Every order of magnitude of frequency increase is anticipated to result in also about one order of magnitude of endurance extension. However, the general problem caused by the internal bias fields remains.

5.2 Polarization Fatigue

Fatigue refers to the gradual degradation of the measurable polarization going to higher numbers of switching cycles.[58, 59, 22] The remanent polarization is obtained by measuring the switching of the polarization. Every domain that is stuck for some reason does not contribute to the macroscopically sensed displaced charge anymore. These domains are also no longer useful for the response of a ferroelectric memory, which depends on a largest possible difference between the two memory states. Every stuck domain reduces this difference.

Fatigue was studied in the same set of experiments used for the wake-up effect in section 5.1. No phase jumps in the harmonics are expected as explained at the end of section 3.7, if the main change in the hysteresis shape is only a polarization decrease as sketched in Fig. 5.1 (or also Fig. 2.6 d)). Thus, the quantity used here to determine a point in time to be plotted vs. temperature in the Arrhenius plots, is the maximum in P_r, which marks the starting point for fatigue. Fig. 5.7 shows the results for frequencies of 100 Hz, 1 kHz and 10 kHz. Activation energies between 48 meV and 103 meV are extracted from the slopes. Similar to the E_A values for the wake-up, they will be discussed later on in section 5.4.

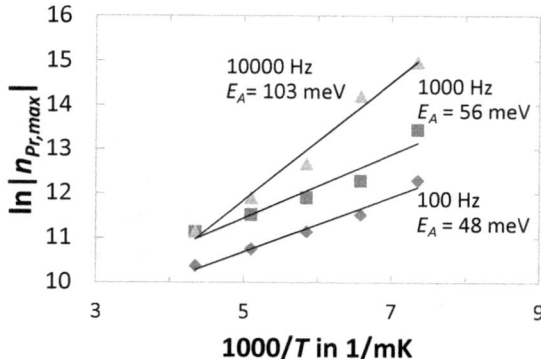

Fig. 5.7: Arrhenius plot for the fatigue at different frequencies for 4 MV/cm field cycling amplitude: Number of cycles $n_{P_{r,max}}$ at which the maximum remanent polarization was observed vs. inverse temperature.[48]

5.2 Polarization Fatigue

FORC measurements were performed at room temperature. To be able of reaching a high enough number of switching cycles to assess the onset of fatigue before dielectric breakdown limits the lifetime of the capacitors at 10 kHz, a field amplitude of 3 MV/cm was chosen. As evident from Fig. 5.8 b) and c), mainly a decrease of the peak amplitude in the switching density is observed. The coercive field stays nearly constant around 2.0 MV/cm, however, a slight bias shift from 0.2 MV/cm to 0.3 MV/cm occurs. In a symmetric structure with symmetric stress, this is not necessarily expected. The 60 FORC curves per measurement point with fixed starting point at positive amplitude should be negligible compared to the amount of switching cycles. However, the observed shift points toward a rising asymmetry in the nominally symmetric TiN-Sr:HfO$_2$-TiN stack.

Fig. 5.8 a) shows the evolution of $2P_r$ and the value of the maximum in the switching density plots ρ_{max} as a function of the number of switching cycles. It reveals another interesting detail: The maximum in P_r is observed at lower cycle numbers than the maximum in switching density. This means that the switching peak still becomes sharper (differences mainly in E_{bias}) while the first domains start not participating in the switching anymore, and a decreased integrated peak intensity is observed. Therefore, it is very likely that different root causes are responsible for the wake-up and fatigue effect in this sample.

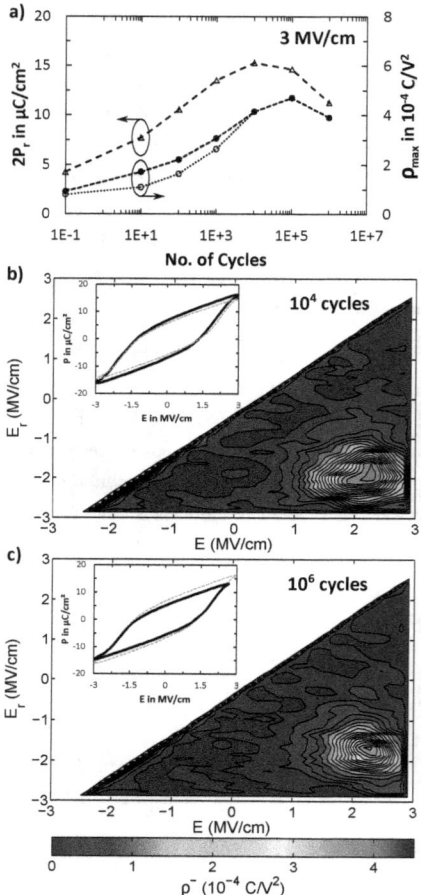

Fig. 5.8: Fatigue for 3 MV/cm amplitude at 10 kHz: a) Evolution of $2P_r$ and the peak height(s) of the switching density ρ_{max} with the number of switching cycles and corresponding switching density plots obtained b) after 10^4 cycles, i.e. after wake-up and c) after 10^6 cycles, i.e., in a fatigued state. The solid line of each inset is the hysteresis for the accompanying switching density plot, whereas the dashed lines are the respective hystereses of the two other cycling stages for comparison.[187]

5.3 Split-up and Merging of Switching Peaks in Transient Currents

Subjecting a sample with an open hysteresis (e.g. after wake-up treatment) to a pulse train of non-saturating field amplitude (subcycling), the peak in the transient current characteristic can be split into two distinct peaks.[48] As Fig. 5.9 shows, the longer the cycling sequence, the more pronounced the split up. The split-up is induced around the field used for the subcycling amplitude (Fig. 5.10). Moreover, even multiple split-ups are possible, if the subcycling sequences are performed in decreasing order of subcycling amplitude. The phenomenon is reversible by applying switching cycles of saturation field amplitude again.

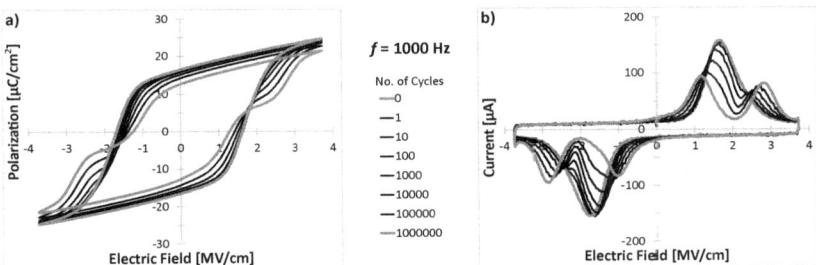

Fig. 5.9: a) Pinching of the P-E hysteresis and b) Split-up of the corresponding currents peaks in I-E graphs with increasing number of cycles with an electric field of 2 MV/cm.[48]

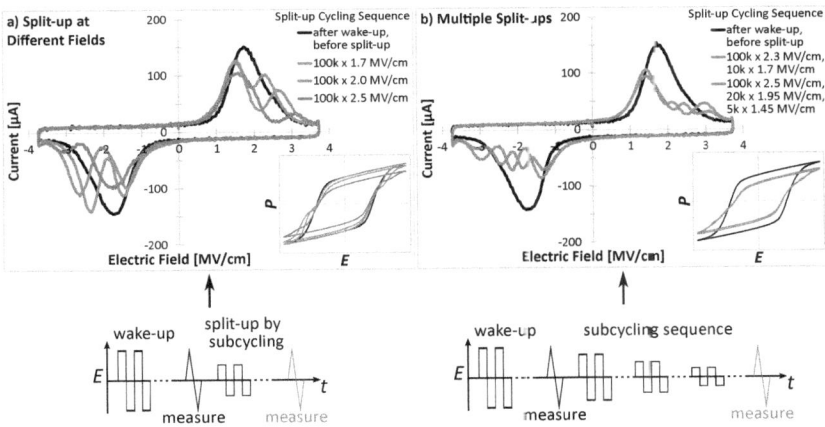

Fig. 5.10: a) Current peak split-ups occur directly at the values of the cycling field amplitude (constant no. of cycles). b) Split-up of multiple peaks by subsequent field cycling with different field amplitudes in descending order. The split-up at 1.45 MV/cm has already vanished in the first quadrant (red line). All experiments were performed with $f = 1000$ Hz. The corresponding P-E hystereses are shown as insets with the E- and P-axes scaled from -4 MV/cm to 4 MV/cm and from $-30\,\mu C/cm^2$ to $30\,\mu C/cm^2$, respectively.[187]

As tentative explanation, the following model approach was proposed in the original article[48]:

- The finite width of the switching current peak is mainly due to different coercive fields. A subcycling field somewhere within the corresponding field range, thus, allows only a fraction of these domains to take part in the continuous switching.
- The switching field of a domain increases with the amount of defects in its volume or at its boundary.[31]
- Defects accumulate at static (non-switching) domains due to their stable mechanical and electrostatic environment.
- Consequently, domains with an initial coercive field above the subcycling field accumulate defects from the domains with an initial coercive field lower than the subcycling field. The former correspond to the higher field peak moving towards higher and higher fields, whereas the latter result in the lower field peak moving more and more towards lower fields.

This model corresponds to the scenario h) shown in Fig. 2.6. A mathematical simulation was also presented and supported this mechanism to account for both wake-up and all features of the split-up/merging process shown until here. However, a later FORC study[187] could provide a new perspective and showed that the original model for this new phenomenon is not the important mechanism behind the split-up/merging effect. Fig. 5.11 a), b) and c) show the respective switching density plots for a single split up, after re-merging the single split-up and for a double split-up (two subcycling fields with decreasing order of amplitudes), respectively. Fig. 5.11 b) shows that the re-merged case looks similar to what was obtained after wake-up in Fig. 5.5, which supports the claim of reversibility of the split-up. Comparing Fig. 5.11 a) to Fig. 5.2, it is obvious that not a split-up in E_c but in E_{bias} direction took place. The switching density looks similar to the situation before wake-up. The manifestation of the double split-up looks even more challenging: four current peaks appear, which not only differ in E_{bias}, but also in E_c.

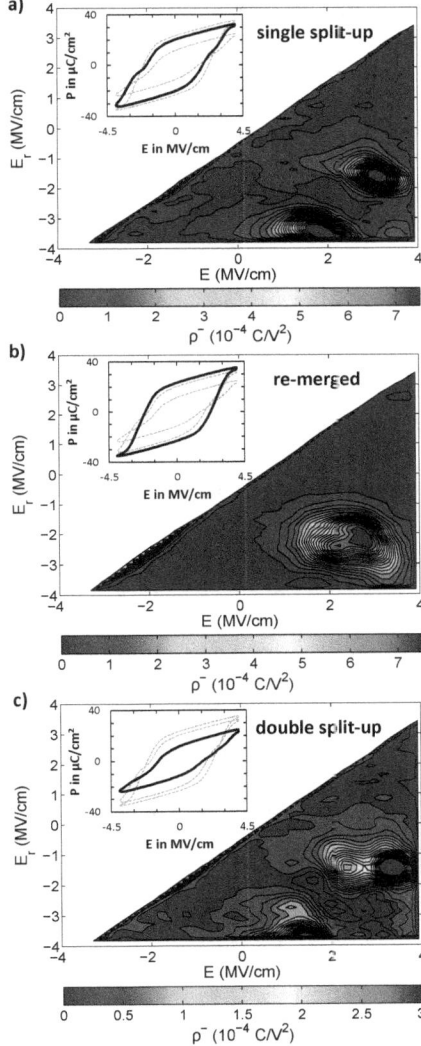

Fig. 5.11: Split-up/merging for 4 MV/cm amplitude after wake-up treatment (see Fig. 5.5 c)): Experimental switching density determined for a) a capacitor subjected to 10^6 cycles of 2 MV/cm (split-up) amplitude, b) the same capacitor after 10^4 merging cycles of 4 MV/cm amplitude, and c) a capacitor subjected to a double split-up by subsequent cycling at 2.5 MV/cm and 1.5 MV/cm for 10^6 cycles each. Note the different pseudocolor-scale of (c). The solid line of each inset is the hysteresis for the accompanying switching density plot, whereas the dashed lines are the respective hystereses of the two other cycling stages for comparison.[187]

It has been shown, that the split-up phenomenon, even with multiple split-ups can be explained solely based on **internal bias fields**.[187] An understanding of the accessible range of switchable domains, as illustrated in Fig. 5.6 together with a consideration of the utilized pulse train of the whole experiment—cycling sequences and FORC measurements—is necessary. The pulse trains are shown in the insets of Fig. 5.12 a) to d).

The following step-by-step explanation refers to the case of the single split-up (Fig. 5.12 c)), but can be applied by analogy to the case of multiple split-ups (Fig. 5.12 d)):

During wake-up cycling (Fig 5.12 a)), all domains are continuously switched as indicated by the white P for polarization and the white $\updownarrow$ for the polarization direction. After the FORC measurement before the subcycling (Fig 5.12 b)), all domains are switched toward positive polarization because the FORC measurement ends with a sweep to positive saturation and back to zero.[7] This situation remains unchanged during the the first positive subcycling pulse (Fig 5.12 c)), which would switch all domains with $E_s \leq E_{sub}$ to positive polarization. Now, during the first negative pulse of the subcycling sequence (Fig 5.12 c)), all domains with $E_{bs} \geq -E_{sub}$ are switched to negative polarization. The rest is stuck at positive polarization (lower part of the Preisach plane with the two white $P \uparrow$ symbols). During the next positive subcycling pulse, only the domains fulfilling $E_s \leq E_{sub}$ are switched back to positive polarization. The rest is stuck at negative polarization (upper right part of the Preisach plane with the white $P \downarrow$ symbol). All other domains, i.e. the ones with both $E_s \leq E_{sub}$ and $E_{bs} \geq -E_{sub}$, are continuously switched during the subcycling sequence (upper left part of the Preisach plane with the white $P \updownarrow$ symbol). The switching domains are subjected to a symmetric voltage stress during the whole subcycling sequence. Thus, they do not get biased into one or the other direction. In contrast, the static domains experience asymmetric conditions because of their fixed polarization. Mobile charges or injected electrons arrange in a way that stabilizes this polarization direction. Thus, they become more and more biased toward the bias direction of opposite sign to their fixed polarization. This explanation is completely analog to what happens during the imprint effect discussed in section 2.4. However, the phenomenon described here, is locally different for different domains and could, therefore, be understood as a **"local imprint"**[58] effect.

---

[7]This is a slightly simplified picture, because all domains with positive backswitching field E_{bs}, i.e. the area above the E_s-axis are switched toward negative polarization. However, virtually no domains with positive E_{bs} exist.

5.3 Split-up and Merging of Switching Peaks in Transient Currents

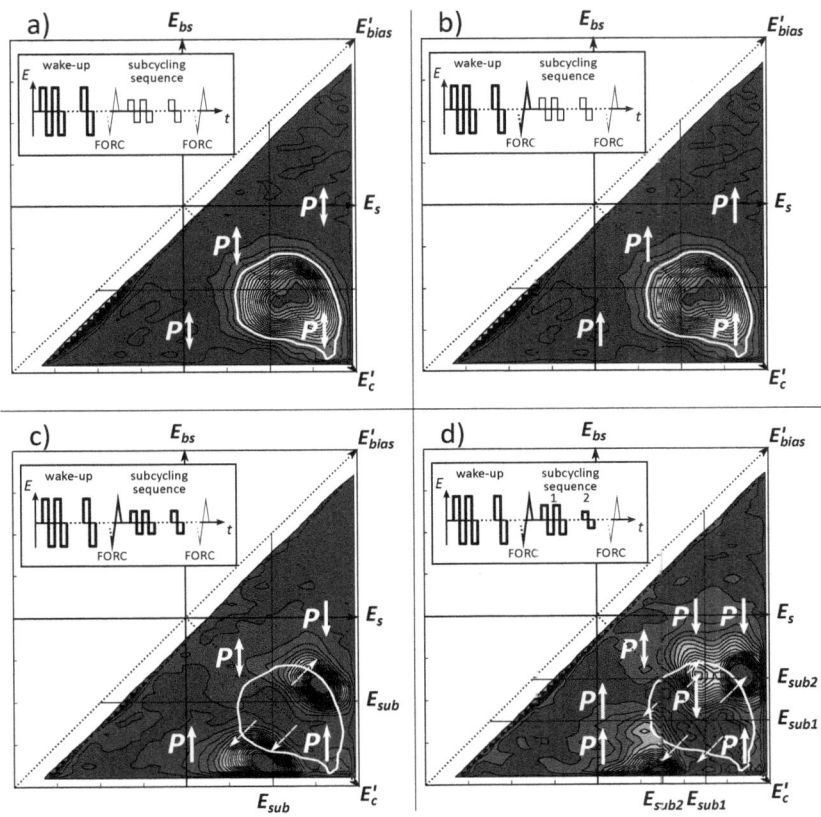

Fig. 5.12: Evolution of the polarization state during the pulse train of the split-up experiments. See text for a detailed explanation of the manifestation of the split-up effect as a local imprint/internal biasing mechanism with a) – c) one and d) two subcycling field(s).[187]

Further proof for this complex interplay of local biasing is provided by two important details in Fig. 5.12 d)[187]: An asymmetry with respect to the diagonal E'_c-axis exists. This asymmetry does not originate from an asymmetric sample configuration. Repeating the whole experiment with interchanged connections to top and bottom electrode results in a completely identical switching density plot. This asymmetry is to be expected from the model because the central square in the switching density plot with $E_{sub2} < E_{s/bs} < E_{sub1}$ is shifted toward the positive E_{bias} direction. An additional consequence of the proposed model is a remaining maximum around zero bias and lowest coercive field, which corresponds to the domains steadily switched during the whole subcycling sequence. In Fig. 5.12 d) and also in Fig. 5.12 c) as well as Fig. 5.11 a) and c), a slight but not significant plateau is observed in the corresponding square. This might also hint at the interactions between the different domains and/or the moved charges and defects. Such interactions are not taken into account in this rather simple model, but are anticipated in real samples.

Finally, it should be noted that the FORC approach is not able to prove any migration of charges, defects or other structural changes. It just shows the frequency of certain switching and backswitching (or bias and coercive fields) assuming that the situation does not change during the measurement itself. A discussion on the mechanisms behind is provided in section 5.6.

5.4 Discussion of Activation Energies

Table 5.1 shows the activation energies extracted from harmonic analysis. Two main trends become clear, that might seem surprising at the first glance:

1. E_A increases with increasing frequency.
2. E_A is also higher for harmonics of higher order.

Moreover, for all three phenomena, wake-up, fatigue, split-up/merging, a dependence on the number of polarization reversals exists, i.e. no pure time-dependence was observed. For different frequencies, the y-intercept changes by less than the expected factor of 10 (offset of ≈ 2.3 on the logarithmized axis). This is in contrast to what was reported by Morozov and Damjanovic for PZT.[38]

Tab. 5.1: Comparison of activation energies for wake-up, fatigue, and remerging after split-up.[70]

Phenomenon[8]	f = 100 Hz	f = 1000 Hz	f = 10 kHz
Wake-up: $E_{A,7}/E_{A,9}$ in meV	84($\pm$6)/88($\pm$7)	98($\pm$20)/99($\pm$24)	105($\pm$12)/126($\pm$22)
Fatigue: E_A in meV	48($\pm$7)	56($\pm$14)	103($\pm$13)
Split-up at 1.9 V: $E_{A,7}/E_{A,9}$ in meV	118($\pm$38)/132($\pm$13)	142($\pm$117)/152($\pm$37)	165($\pm$37)/185($\pm$26)
Split-up at 2.3 V: $E_{A,7}/E_{A,9}$ in meV	105($\pm$29)/129($\pm$22)	111($\pm$40)/129($\pm$28)	124($\pm$28)/138($\pm$35)

To understand these trends, it needs to be considered that the series of harmonics is only a mathematical representation of the measured polarization hystereses. As can be seen from the example P-E data in Fig. 5.13, the switching peaks become broader and move towards higher fields for higher frequencies. This results in a decrease of the measured polarization values because the electric field is not high enough to sufficiently saturate the hystereses. Higher fields resulted in a too early breakdown of the samples. This limitation results in a lowered amount of domains that take part in the switching process at higher frequencies. The domains with higher switching fields are missing. The effective average over all participating sample regions is thus expected to differ. Regarding the direction of the difference, the

5.4 Discussion of Activation Energies

origin of domains with higher switching fields has to be considered. As Fig. 5.6 shows, two differently biased regions exist. Only the switching toward one direction in each of these two regions is impaired by a slight movement of the switching peaks toward higher fields. This direction is the more critical one for the polarization reversal process, which was shown to be important for the defect related processes that underlie wake-up fatigue and the split-up/merging phenomenon. The difference between the maximum amplitude and the highest switching field is lower for higher frequencies. The redistribution process, thus, might depend to a greater extent on thermal activation. The effective band bending to overcome local energy minima is lower at 10 kHz compared to 1 kHz or 100 Hz. However, Morozov and Damjanovic reported higher activation energies for higher amplitudes used for wake-up cycling.[38, 190] The P-E hysteresis shapes were not shown and discussed. The plotted evolutions of the third harmonic allow to conclude that the shape was also affected for the studied PZT samples.

Interestingly, a frequency dependence of activation energies similar to the ones for fatigue was also observed in reliability studies of dielectric break down in HfO_2 based transistor gate stacks.[269] It is well-established that dielectric breakdown in these gate stacks is due to oxygen vacancies generated near the Si-O(-N) interface layer to the transistor channel.[270]

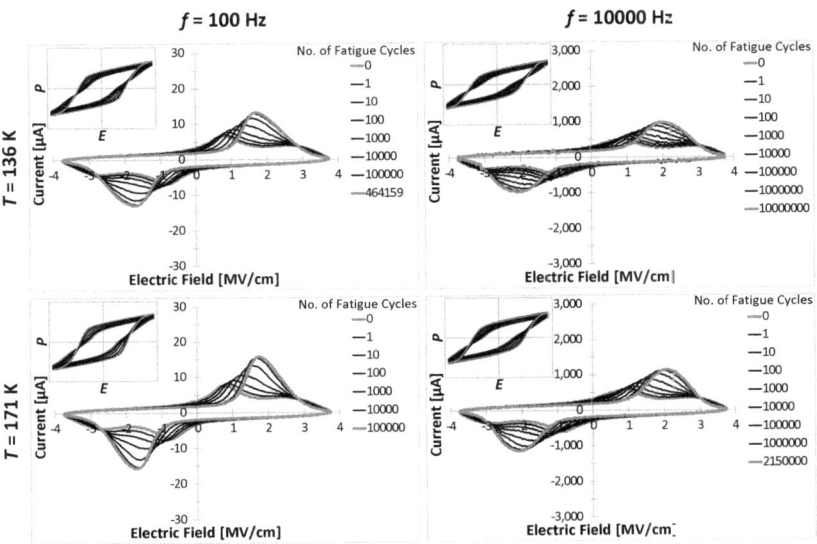

Fig. 5.13: Evolution of the hysteresis shape during wake-up for 100 Hz and 10000 Hz. For higher frequencies, lower P_r values are observed and the hysteresis edges are less steep.

After having discussed the frequency dependence, the question why higher values are obtained for higher harmonics has to be settled. Phase jumps in higher harmonics were observed later in time. If multiple phenomena are involved in the field cycling behavior, some might dominate at the beginning and others more toward the end of the cycling sequence. Without deeper knowledge of the underlying mechanisms, this assumption can at least explain the observation of different values in E_A.

For the split-up at 2.3 MV/cm a lower activation energy than for the split-up at 1.9 MV/cm subcycling amplitude was obtained. Looking at Fig. 5.12, it is obvious that for higher subcycling voltages, a lower amount of domains is static and thus, subject to local imprint. On average, these domains have a higher coercive field. Moreover, a higher band bending is caused by the higher subcycling field, which could fill deeper traps or move oxygen vacancies across higher local barriers. This would suggest a higher activation energy for the split-up at 2.3 MV/cm. On the contrary, as domains become very strongly biased, either switching or backswitching fields become too high for the subsequent remerging amplitude. Thus, the most strongly biased domains do not take part in switching during the merging process. The lower E_A observed for the higher subcycling field suggests the latter effect might be the dominant one here. However, the interplay might be of a more complex nature.

The obtained energies for wake-up (aging/deaging) or fatigue are nearly one order of magnitude lower than the values around 0.6 – 1.2 eV reported for $BaTiO_3$[271] or PZT[38, 272, 41]. Moreover, a different temperature range had to be chosen to assess the mechanisms on a reasonable time scale.

Pinched hystereses due to internal bias fields were discussed in section 2.4 (Fig. 2.6 g)). So-called "grain boundary", "domain wall" and "volume effects" are distinguished in perovskites. The **grain boundary effect** represents space charge accumulating in a second phase around the ferroelectric grains as it was shown for Al doped PZT[41]. This hypothesis could be varified e.g. by TEM methods or IS (section 5.6). A movement of dopant ions is the base of the **domain wall effect**: After aging, the dopant ions are inhomogeneously distributed and a diffusion to domain walls and exchange of electrons between different valance states could have taken place. Finally, the **volume effect** describes the alignment of mechanical and electrical dipoles (anisotropic centers) in favor of the polarization state present in the respective domain. However, the domain wall and volume effect are hard to discriminate and both were reported[41, 38]. In PZT, the dipoles reorienting within a unit cell result form doping with acceptor ions, such as Fe.[268] Acceptor doping induces oxygen vacancies to preserve charge neutrality in ceramics[273]. In thin films, this is achieved by holes rather than O vacancies.[274] Dopant contents can be in a similar range as for the hafnia thin films used in this work. However, hafnia and zirconia thin films deposited by ALD or PVD are usually substoichiometric even without doping.[275] Moreover, oxygen is the switching ion in the hafnia unit cell (Fig. 2.11) whereas it is part of the octahedron surrounding the switching

5.4 Discussion of Activation Energies

Ti and Zr ions in the perovskite cell (Fig. 2.3). This means, there are some considerable differences between PZT and hafnia/zirconia based ferroelectrics.

The problem of polarization fatigue in commercial PZT based memories was solved by, firstly, carefully adjusting the Pb stoichiometry[276, 277, 59] and, secondly, the use of Ir or IrO_2 electrodes[22], that do not scavenge oxygen from the PZT films. The used TiN electrodes are known to partially oxidize under anneal conditions comparable to the ones used in this work.[278, 279, 200] Thus, they could play a role for both wake-up and fatigue. Hafnia (resistive switching[127]) and its sister-oxide zirconia (fuel cells[125, 126]) are both known to be good oxygen conductors.[116, 117] Recently, resistive switching was shown for films that also exhibited ferroelectric properties.[280] Moreover, an in-situ TEM study on similar hafnia films under similar condition confirmed the generation of oxygen ions and vacancies.[281] Finally, simulations also point toward the role of O ions and vacancies.[282]. Now, the energy values for these processes need to be discussed: The barrier for single jumps of vacancies in a monoclinic lattice can be as low as 50 meV, whereas overall effective barriers of 0.7 eV are anticipated for double positively charged and even 2.4 eV for neutral O vacancies.[283] Value of 0.7 eV – 1.5 eV below the conduction band are also found as a typical electronic trap level in zirconia and hafnia films for state-of-the-art CMOS transistors[284] and DRAM capacitor stacks[131]. With a work function of ≈ 4.9 eV, a band gap of ≈ 5.5 eV – 6 eV and an electron affinity of ≈ 2.2 eV, these defects align between ≈ 2.1 eV – 0.9 eV above the Fermi level of the TiN electrode. These states that are of special interest for electronic conduction via trap-assisted tunneling. Besides those states, shallow traps due to oxygen vacancies are found between 0.3 eV and 0.5 eV below the conduction band. In case of sufficient electron injection from the electrode, these trap levels would be of most interest for electronic conduction via Poole-Frenkel emission.[285]

Besides the role of oxygen in the unit cell, another main difference to the perovskite ferroelectrics are the at least one order of magnitude higher fields used for switching. Instead of 10 – 100 kV/cm, a few MV/cm are needed for hafnia and zirconia based ferroelectrics. Moreover, grains and consequently also grain boundaries span the whole film thickness as shown in Fig. 4.14. A PFM study on Si:HfO_2 claimed similarly poled regions to consist of several grains[243]. I.e. each of these regions could potentially contain ion and electron conducting paths.

However, remembering the findings of section 4.5, the situation becomes more complex: Oxygen vacancies also influence phase stability. Thus, their movement and redistribution can also alter phase stability of certain regions. Values of barriers between different phases/polarization states are expected to be between 30 and 600 meV as summarized in section 4.1. But these values only reflect the phase transitions themselves and not the assisting effect of an oxygen vacancy redistribution. In a measured effective activation energy, mainly the limiting process will be reflected.

Comparing energy values from literature to the ones extracted in this work can be misleading. The values given in Tab. 5.1 do not correspond to a certain barrier calculated from ab-initio at zero field. To asses, which mechanism accounts for these phenomena a simulation with applied AC field is necessary to calculate such an effective barrier. However, since an electric field is anticipated to lower the present barriers, they can serve as a lower limit. I.e. any phenomenon with a zero-field activation energy below these values cannot be the one governing the effects observed here.

DC experiments have been described elsewhere[130]. They could be an easier way to assess at least that part of the underlying effects of field cycling behavior, that do not require a continuous switching treatment.

To conclude this section, a **remark on the use of different extraction methods** for the activation energy should be given as an evaluation of pros and cons for future work in this direction:

1. Similar to this work, Morozov and Damjanovic[38] used **phase jumps in higher harmonics** to assess the wake-up phenomenon. The applicability and drawbacks of this method have been discussed above.
2. Also for wake-up, Carl and Hardtl used the diminishing of **internal bias fields** as defined by the difference in the corresponding positive and negative switching field fields of merging peaks in transient currents. This criterion does not account for an increase in the amount of domains taking part in the switching or a reorientation of them toward the applied field.
3. A more precise method would be **FORC measurements** at different cycling stages and for different temperatures to evaluate the evolution of the internal bias fields/and or E_c and P_r. However, the FORC measurements represent an additional (asymmetric) cycling of the sample, which is a perturbing effect especially for the wake-up taking place at lower cycle numbers.
4. The **start and end of a stable plateau of a maximum** P_r value could be used to mark the end of the wake-up and the start of fatigue, respectively. If no such plateau of a stabile P_r is present, i.e. wake-up and fatigue strongly interfere, a clean extraction is hampered.

5. Using the time until a certain **fixed** P_r value is reached, is not a good criterion given the temperature dependence of the polarization presented in section 2.2. For the same reasons, also the use of **fixed switching fields** is not recommended.
6. The time until the maximum in P_r is observed or until P_r has increased by a **certain ratio compared to the initial** P_r could be used to extract an activation energy for the wake-up. The other way around: Starting from a maximum in P_r the time to a drop by a certain percentage of that maximum could be used for fatigue.
7. If a certain **model of wake-up or fatigue** exists, that reasonably predicts the course of the P_r or bias field vs. time/cycle number, slopes or some other key parameter for the speed of the underlying process could be extracted from a **fit** and plotted vs. $1/T$. This was not the case in this work, but could be subject of future studies.

As will be shown later, the vanishing of bias fields and increase of P_r during wake-up might be due to two different effects. Therefore, it could be useful to treat them separately, e.g. by evaluating bias fields and P_r. However, a certain interference of the phenomena cannot be excluded as a saturation of the hystereses is limited by dielectric breakdown.

Summing up, a single obviously most practicable and least error-prone method is not easily at hand. The temperature dependence of the double-well potential interferes with several parameters, that are related to wake-up and fatigue. Thus, using different extraction methods on one set of experiments can easily give strongly different values. Depending on the specific behavior of their material system, researchers are encouraged to choose the most proper method according to the remarks given above.

5.5 PFM to Study Switching on Nanoscale

Piezoresponse force microscopy (PFM) was applied to study the domain evolution on the nanoscale. All data shown below were obtained during a research visit at the Oak Ridge National Laboratories in the frame of the user program of the **Center for Nanophase Materials Sciences (CNMS)**. A description of the sample preparation and the measurement methods can be found in sections 3.1 and 3.3, respectively. All measurements below were performed on the following sample: 27 nm thick Gd:HfO$_2$ with TiN top and bottom electrode annealed at 650 °C for 20 s.

A single-frequency PFM (SF-PFM) image is shown in Fig. 5.14. The measurements were obtained with a metal coated tip (BudgetSensors ElectriMulti 75E) on bare oxide. A clear PFM-like phase and amplitude signal is present on the whole scan area. As can bee seen, an enhancement of the piezoresponse at the grain boundaries is obvious, which hints at geometry effects between tip and surface (so called "crosstalk").

| Topography | Amplitude | Phase Angle |

a) preliminary scan (different spot)

b) scan before BEPS and cKPFM (same spot as c))

c) resonance curve with V_{AC} = 1 V before SF-PFM, BEPS and cKPFM (same spot as b))

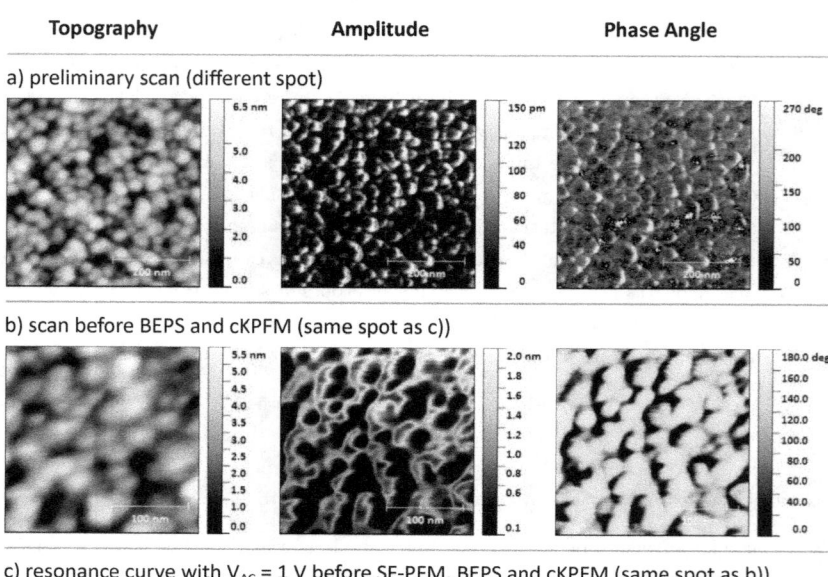

Fig. 5.14: SF-PFM results on bare oxide for a 27 nm thick Gd:HfO$_2$ sample a) in a preliminary experiment and b) before BEPS and cKPFM and with the same tip in the same spot as these subsequent measurements. c) The frequency sweep prior to the series of SF-PFM scans, BEPS and cKPFM measurements: A good signal-to-noise ratio is observed. Frequency sweeps after the measurement sequence usually exhibited amplitudes that were at least a factor of ten lower.

5.5 PFM to Study Switching on Nanoscale

Band excitation point spectroscopy (BEPS, see section 3.3) is capable of measuring pixelwise a local hysteresis of the piezoelectric response. Fig. 5.15 shows the results for the 27 nm thick Gd:HfO$_2$ sample. At 14 V and -14 V bias, the phase angle φ of nearly the whole 250 nm × 250 nm area is 140° and $-40°$, respectively. The difference of 180° matches the expectations for a ferroelectric. Interestingly, the overall amplitude at 14 V is lower than for -14 V, which indicates an overall bias towards negative voltages. Two differently biased regions exist as is exemplarily indicated with A and B in Fig. 5.15: Regions similar to A are already switched to positive polarization at 8.75 V (red spots in the phase map), whereas regions like B still remain in the negative state and some do not even switch at $+14$,V (white and blue spots in the phase map). At -1.75 V, all B-like regions are switched back to negative polarization again (blue spots in the phase map), whereas A-like regions just start to switch back (white and red spots). A white pixel indicates that no successful SHO fit was possible, because the piezoresponse is zero (or nearly zero) around the switching fields. The observation of two differently biased regions on microscale appears to be similar to what was obtained for the pristine capacitor samples in section 5.1 via macroscopic FORC measurements.

Finally, looking at the resonance frequency ω_0 and quality factor Q (compare Fig. 3.7), it can be seen that those are voltage independent and only result from the tip-surface contact (topography). The values of ω_0 span about 5...6 kHz across the scanned area. In SF-PFM around resonance frequency, this would already result in fluctuations in both phase and amplitude highlighting the advantage of the BE approach. Interestingly, there seems to be a correlation between a lower ω_0 and a lower value in Q. A lower resonance frequency can be achieved by a lower effective force on the tip, which might give rise to worse coupling of the electrical excitation by a reduced contact area.

Balke et al.[286] demonstrated an artificial piezoresponse in similar order of magnitude can be measured in amorphous HfO$_2$. Ionic motion, charge injection and electrochemistry were some of the listed effects to account for a pseudo-PFM response. In case of measurements directly on the dielectric surface, electrostatic contributions are inevitable. Despite the fact that the obtained results align well with the FORC results of sections 5.1 – 5.3, this crosscheck is mandated to prove whether the results are clean or not. As a first step, they recommended to check both on- and off-field loops. Fig. 5.15 shows the comparison of the Gd:HfO$_2$ sample to what was obtained for PZT and amorphous HfO$_2$ in similar measurements. The results for the Gd:HfO$_2$ sample in this work look similar to the curves recorded on amorphous, i.e. non-FE HfO$_2$ by Balke et al.[286]. Consequently, the above interpretation of Fig. 5.15 should be treated with caution. A check for artifacts is urgently needed also in future work on similar films.

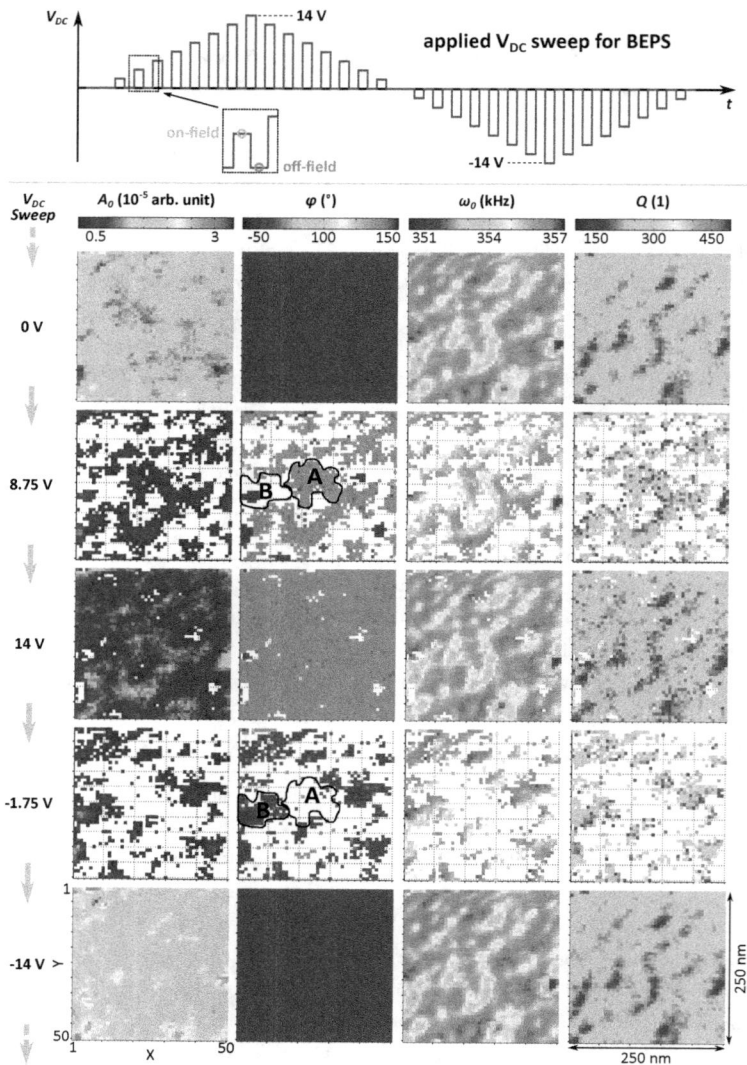

Fig. 5.15: Off-field BEPS on bare HfO$_2$: Evolution of amplitude A_0, phase angle φ, contact resonance frequency ω_0 and quality factor Q during the sketched bias voltage sweep as obtained from the SHO fit of the response in frequency domain (after Fourier transform). At 0 V only one phase and varying amplitudes are observed. The samples has a preferential polarization orientation across the whole probes area. At 8.75 V, part of the domains are switched to opposite polarization (phase changed be 180°, now red color). White pixels denote that no successfull SHO fit was possible due to the low signal around the switching field. At 14 V nearly the whole area is switched to the opposite state. Backswitching to the initial state (blue color in the phase diagrams) starts at around -1.75 V, i.e. at lower absolute voltage values than the opposite switching. At -14 V, the whole sample area is in the initial polarization state. A and B denote examples of the two types of differently biased regions found in the sample.

5.5 PFM to Study Switching on Nanoscale

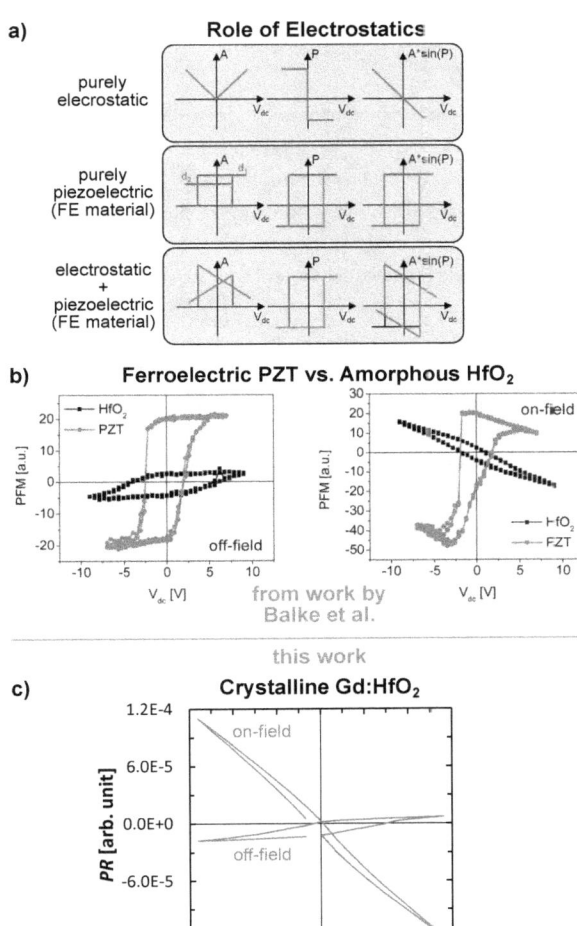

Fig. 5.16: Electrostatic contributions to the piezoresponse (common symbols in publications: PR, PFM, $A \cdot \sin(P)$): a) sketch of the interaction of purely electrostatic and purely piezoelectric signals[166], b) comparison of on- and off-field loops for PZT and amorphous HfO_2[286] and c) averaged on- and off-field loops extracted from 3EPS measurements in Fig. 5.15.

Three criteria to check for a truly FE piezoresponse vs. artifacts were listed by Balke et al. and should be considered here[286]:

1. Contact-mode Kelvin-probe force microscopy (cKPFM) was used to determined the surface potential and to check whether a typical manifestation of the piezoresponse can be identified or not. Fig. 5.17 shows the results of cKPFM. The expected change in the slopes of the cKPFM-signal (similar to the piezoresponse, but not necessarily of same origin) vs. V_{read} curves around the coercive fields is barely visible. The resulting curve looks rather like the one obtained for amorphous HfO_2 than the one measured on PZT by Balke et al.

2. Following Strelcov et al.[287], both on- and off-field PFM loops (normalized to V_{AC}) should collapse if the AC excitation fields are increased and become similar to the coercive field. The former is anticipated to reduce in width and the latter mainly loses height of the PFM hysteresis for increasing V_{AC}. Fig. 5.18 again shows a comparison by Balke et al. for PZT and amorphous hafnia. The results for 27 nm thick HfO_2 in this work show some intermediate behavior: The loops show the anticipated collapse. However, the on-field loop shows very high electrostatic contributions and is 180° phase-shifted (mirrored at the x-axis) compared to the one for PZT and thus, has also some similarities with the nearly straigth lines measured on amorphous HfO_2. Estimated d_{33}-values (strain response and electric field along the [001] direction of the crystal) in the order of 3 pm/V or 10 pm/V from simulation data (unpublished work by A. Kersch's group, Munich UAS) and macroscopic interferometry data[1, 194], respectively, promote the picture that rather low piezoelectric compared to electrostatic contributions explain the problems.

3. A higher cantilever stiffness reduces the electrostatic contributions. With a tenfold stiffer cantilever no response remained on the amorphous HfO_2 of Balke et al. Unfortunately, a similar experiment could not be performed within this work. Moreover, it should be mentioned, that the use of cantilevers with increased stiffness is not straightforward in this where measurements need to be performed at contact resonance frequency. Stiffer cantilever result in a stiffer contact and modifying the cantilever shape at first eigen-mode. Therefore, results for different cantilevers cannot be compared.

The BEPS results are in accordance with the FORC results, but perturbing effects of a magnitude similar to that of the pure PFM response are clearly present.

5.5 PFM to Study Switching on Nanoscale

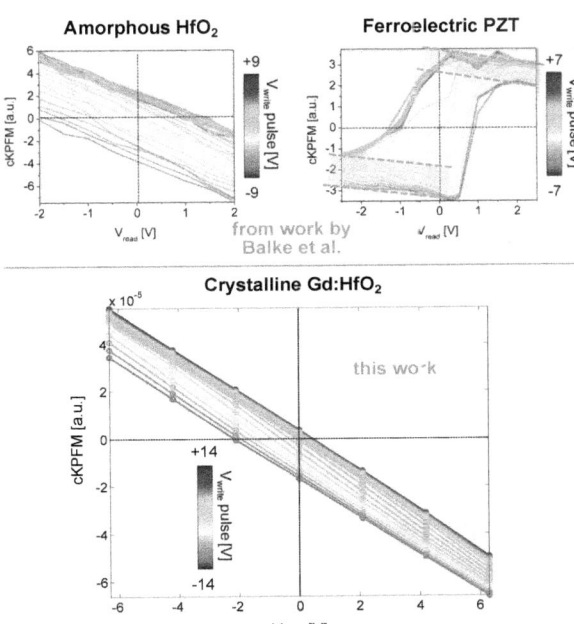

Fig. 5.17: Comparison of cKPFM results for the 27 nm thick Gd:HfO$_2$ film of this work to what Balke et al. showed for a ferroelectric PZT film and an amorphous and therefore non-ferroelectric HfO$_2$ film.[286]

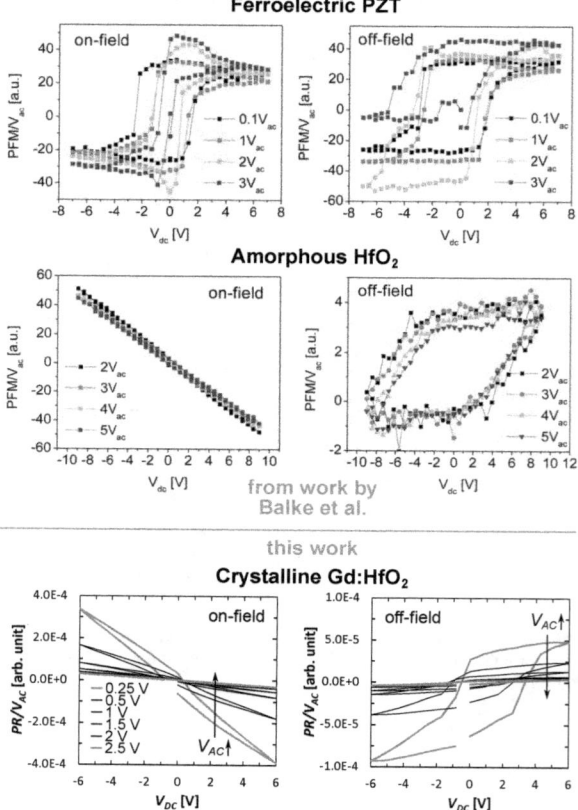

Fig. 5.18: On- and off-field loops of the piezoresponse $PR = A_0 \cdot \cos(\varphi + \Delta\varphi)$ (phase offset $\Delta\varphi$ chosen to maximize loop area) from BEPS averaged over the 50 × 50 pixels in Fig. 5.15 compared to the results for PZT and amorphous HfO$_2$ by Balke et al.[286]

To enable PFM measurements of materials with small-d_{33} in ultrathin films (≈ 10 nm and below) in the future, suitable measurement concepts need to be developed. Directions to follow could include:

- Measurements on capacitors: With the tip on an electrically connected top electrode, the electrostatic forces between tip and sample are reduced. However, roughness of the electrode and a clamping effect on the FE below need to be considered.
- Deeper understanding of contact dynamics: A more detailed understanding helps to separate electrostatic from piezoelectric signal contributions. The approach of cKPFM is one first step in this direction.

Based on the results presented in this section, caution is advised when interpreting PFM results in general and especially for ultrathin films of hafnia/zirconia based ferroelectrics. E.g. a recent study on 2.5 nm (a tenth of the thickness of the films used in this work) thin $Hf_{0.5}Zr_{0.5}O_2$ films[288] contains 1) phase jumps of less than 180°, 2) non-saturating butterfly curves of the PFM amplitude and 3) no signs of any non-uniformity in the distribution o ferroelectric PFM properties despite the low P_r of only 3µC/cm^2 that was measured. Moreover, the crystallization temperature strongly increases for such ultrathin films. SEM and TEM data was shown to prove the crystalinity of the films. However, remaining amorphous fractions would not be surprising.

5.6 Structural Evolution During Field Cycling

In order to assess structural changes causing the evolution in the switching density found in sections 5.1 to 5.3, impedance spectroscopy (see section 3.5) and TEM methods (see section 3.4) were applied.[185] The same sample as for the identification of the FE phase in section 4.2 was used: 27 nm thick Gd:HfO$_2$ with TiN top and bottom electrode, annealed at 650 °C for 20 s.[171]

Hints at structural changes were found in small-signal C-E measurements as shown in Fig. 5.19 a) and b). As anticipated from the FORC results, the depinching of the P-E hysteresis during wake-up is accompanied by a sharpening of the maxima in the butterfly-shaped curve of k vs. bias voltage, which stem from domain wall contributions[161]. When fatigue sets on (decrease of P_r), these maxima also decrease. However, there is a continuous trend of the minimum relative permittivity k_{min}, which is representative of the crystallographic phase, decreasing with cycling. The pristine sample possesses values closer to what has been reported for the higher symmetry tetragonal or cubic phases, whereas during fatigue, k_{min} becomes closer to the reported values of the monoclinic hafnia. The range of k-values reported in literature are given in Tab. 5.2 and the corresponding spans are included in Fig. 5.19 a). A bright field TEM image of the studied stack of the Gd:HfO$_2$ sample is given in Fig. 5.19 c).

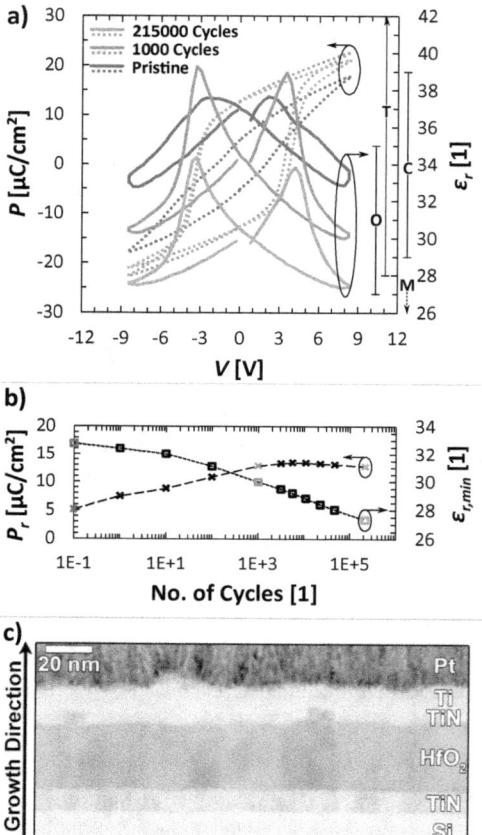

Fig. 5.19: a) Hystereses of polarization P (dotted lines) and relative permittivity k (solid lines) obtained from small-signal capacitance measurements for the three different regimes: pristine (0 cycles), during wake-up (1000 cycles) and during fatigue (215000 cycles); b) the trend of remnant polarization P_r and minimum relative permittivity $k_{r,min}$ vs. number of switching cycles; c) a bright field STEM image showing the whole film stack and HfO$_2$ grains that span the whole film thickness.[185]

Tab. 5.2: Relative permittivities from literature for the ferroelectric orthorhombic Pca2$_1$[61, 132] and the paraelectric monoclinic P2$_1$/c[61, 132, 133, 134, 135], tetragonal P4$_2$/nmc[61, 133, 135], and cubic Fm3m[61, 132, 133, 134, 135] phases.[185]

Phase	Space Group	ε_r [1]
C	Fm3m	ε_r = 29...39
T	P4$_2$/nmc	ε_r = 28...70
O (FE)	Pca2$_1$	ε_r = 27...35
M	P2$_1$/c	ε_r = 16...20

5.6 Structural Evolution During Field Cycling

Impedance spectroscopy was used to gain more insight into the changes of the dielectric properties of the sample. In fact, it is necessary to prove that any electrical small-signal response at a chosen frequency is dominated by the ferroelectric bulk properties and not by interfaces or grain boundaries. This is seldom checked or mentioned in literature despite its importance for temperature dependent measurement of the relative permittivity to identify phase transitions. A very basic equivalent circuit of an RC combination was discussed in section 3.5. This circuit accounts for the dielectric properties and the finite conductivity of the HfO_2 and an additional series resistance representing non-ideal electrodes or uncompensated contact/cable resistances. In contrast to ceramics of thicknesses in the mm range, grain boundaries are not anticipated to give rise to a second RC element in series to the first one. Grains and therefore also the grain boundaries commonly span the whole film thickness as can be seen from Fig. 4.14 a) (dark field TEM image) and also from Fig. 5.19 c) (bright field STEM image) below. A series arrangement of the second RC element would be a more suitable reflection of the physical reality. However, such an arrangement is not useful from a mathematical point of view. The system would be underdetermined and the two parallel R's as well as the two parallel C's could always be merged into one R and one C, respectively. Consequently, the modeling strategy does not change for thin films compared to ceramics and mainly series arrangements of RC elements are added until the measured impedance spectra are sufficiently fitted. Later on they are assigned a physical meaning in all conscience. Differences in bulk or interfacial dielectric and conductive properties are scenarios necessitating a series arrangement of two or more RC elements.

The sample studied in this work requires three RC-like elements in addition to a series resistance R_4 to be properly represented by the equivalent circuit. Fig. 5.20 shows the measurement results in different types of Bode and Nyquist plots. The utilized equivalent circuit elements and their counterparts in the film stack are shown in Fig. 5.21. Table 5.3 lists the obtained fit parameters for all circuit elements. The first RC element consists of a linear capacitance C_1 and a resistance R_1, which is too high to be reliably fitted in the accessible frequency range ($f < 40\,\text{Hz}$ would be needed). Because of the high resistance, this RC combination is likely to represent the bulk of the ferroelectric film. A relative permittivity of around 30 can be calculated from $C_1 \approx 1\,\text{nF}$. A resistance R_2 of $10^5...10^7\,\Omega$ and a constant-phase element (CPE) Q_2 form the second RC-like element. The value of R_2, which is orders of magnitude higher than the $100\,\Omega$ of the series resistance R_4, and the CPE-character suggest that this element arises due to a thinner dielectric layer. Literature provides a wealth of potential explanations for CPE-behavior including roughness[183] and permittivity or resistivity inhomogeneities (distributions parallel and/or normal to the electrodes)[289, 290, 291, 292, 293, 294], which would all serve as reasonable assumptions for the present case. For the CPE, a similar permittivity value as for C_1 can be calculated at 10 kHz frequency. A third RC element with a resistance R_3 on the order of the series resistance R_4 and a linear capacitance C_3 is found. Because of the low resistance, it is assumed that this RC element represents a non-ideal electrode rather than a layer of dielectric origin. An interface of

TiO$_x$N$_y$ as it has been reported by Weinreich et al.[278] for ZrO$_2$ film stacks as a consequence of the deposition and annealing procedure would be one explanation. However, no CPE-behavior was needed to model this layer as it would be anticipated for an inhomogeneous interfacial layer with a high ratio of roughness to thickness. Results for Hf$_{1-x}$Zr$_x$O$_2$ by Park et al.[279, 200] suggest that also the complete TiN electrode could be subject to a homogeneous partial oxidation. Assuming a thickness of $\approx$ 1 nm for an interfacial layer would result in a relative permittivity of $\approx$ 6 for all three regimes of field cycling. An interfacial layer of TiO$_x$N$_y$ with such low thickness and permittivity, but still a very high uniformity seems rather unlikely. If a partially but uniformly oxidized Ti or TiN layer of 10 nm or 20 nm thickness is assumed, the absence of a CPE could be explained and the resulting permittivity between 60 and 30 would fall into the range of what was reported for TiO$_2$[128].

Evaluating the fit parameters of the equivalent circuit, the following main trends with electric field cycling can be derived:

1. There is no substantial change in the resistance of the electrode-like layer (R_3, C_3, R_4). However, a drop of the capacitance C_3 occurs during the wake-up stage. No CPE was needed to fit the impedance spectra.
2. During wake-up, the bulk capacitance C_1 remains nearly constant, but drops by 20 % during fatigue. Arguing with these values is problematic because domain-wall contributions are superimposed to the pure phase contribution at zero volts (Fig. 5.19 b)). IS is not capable of monitoring the evolution of the exceedingly large R_1. Nonetheless, from static leakage current measurements (Fig. 5.22), the resistance degradation of this thickest layer during fatigue can be concluded to be at least orders of magnitude .
3. The dielectric interfacial layer (R_2, Q_2) is subject to notable changes of both dielectric and resistive properties. Its resistance (both the R_2 and the real Q_2 contribution) increases during wake-up hinting at a decreasing number of electrically effective defects. During fatigue, it drops by two orders of magnitude. Interestingly, the exponent n of the CPE increases continuously from about 0.87 via 0.91 to 0.93, which corresponds to a decrease in inhomogeneity.

Besides the insights in to the constitution of the current sample, these results highlight the importance of a point stressed at the beginning: A broad frequency spectrum needs to be checked and modeled to be sure that the measurement frequency for small-signal capacitance measurements actually reflects the ferroelectric bulk properties. As can be seen from Fig. 5.20, this is not the case for the frequency of 10 kHz used above. In this range, R_2 and Q_2 already dominate the spectrum.

5.6 Structural Evolution During Field Cycling

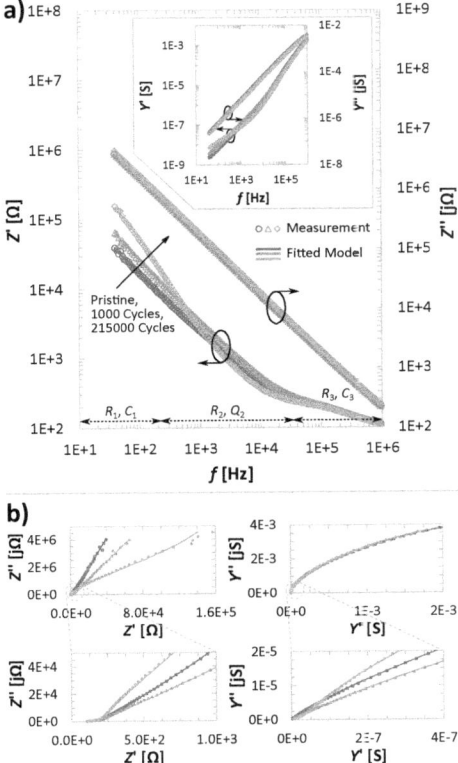

Fig. 5.20: Impedance spectroscopy: a) logarithmic Bode plots of impedance Z and admittance Y (inset); b) Nyquist plots of impedance and admittance in both full scale and magnified sections to show the high quality of the fit throughout the whole frequency spectrum. Blue, green and orange color represent the capacitors in pristine state, after 1000 cycles and after 215000 cycles at 1 kHz and 8.5 V, respectively[185]

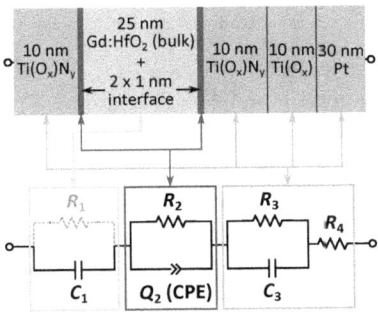

Fig. 5.21: Impedance spectroscopy: Film stack with corresponding equivalent circuit. Fit parameters can be found in Table 5.3[185]

Tab. 5.3: Impedance spectroscopy: Model parameters of the equivalent circuits sketched in Fig. 5.21 used to fit the data presented in Fig. 5.20. Relative standard deviations of all fit parameters are 10^{-4} or below.[185]

No. of Cycles	0 (Pristine)	1000 (Woken-up)	215000 (Fatigued)
R_1 [Ω]	∞	∞	∞
C_1 [F]	1.03E-9	1.10E-9	0.90E-9
R_2 [Ω]	1.06E+7	2.48E+7	2.41E+5
Q_2 [j^nΩ]	T = 4.81E-8 n = 0.8660	T = 1.68E-8 n = 0.9056	T = 2.04E-8 n = 0.9286
R_3 [Ω]	72.9	67.6	78.8
C_3 [F]	5.55E-9	4.69E-9	5.09E-9
R_4 [Ω]	92.5	87.6	94.9

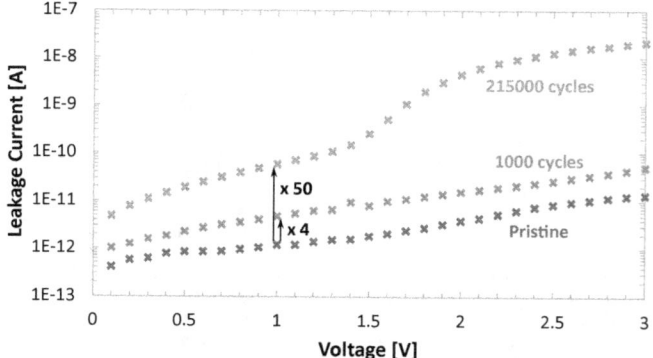

Fig. 5.22: Leakage current characteristics for capacitors in a pristine state (blue), after 1000 (toward the end of wake-up, orange) and after 215000 cycles (during fatigue, green). The arrows exemplarily indicate by what factor the leakage current of the pristine sample at 1 V multiplied during wake-up and fatigue.[185]

Aberration corrected STEM was conducted to help gaining insight into the structural changes and to potentially correlate them to the evolution of the electrical properties. Multiple regions were studied via HAADF-RevSTEM in samples subjected to the same field cycling treatment as the ones used for IS to identify the phases of these regions. Lower magnification images were later on used to assess the size of these grains. In all three samples only monoclinic and orthorhombic grains were found. Nine, nine and eight regions were suitable for phase identification in the samples in pristine, woken-up and fatigued state, respectively. The orthorhombic phase fractions were found to increase from 10 % to 80 % to 90 % with increasing numbers of cycles. The remaining regions are all orthorhombic. Estimated uncertainties of

5.6 Structural Evolution During Field Cycling

±10 % were due to limited STEM sampling and challenges in area measurements due to localized tilting, grain overlap, etc. Therefore, the composition of the woken-up and fatigued states are within error of each other. Table 5.4 summarizes the aforementioned findings for the change in phase composition. In contrast to what was described in section 4.2, the orthorhombic phases were not individually checked for polarity. However, from those results it is safe to infer that many of the orthorhombic grains were also polar, i.e. consisted of the ferroelectric $Pca2_1$ phase. Moreover, the P_r values increased with cycling as shown in Fig. 5.19 b) and c).

Tab. 5.4: Tabulated monoclinic (M) and orthorhombic (O) phase fractions measured from the pristine, woken-up, and fatigued samples together with corresponding remanent/maximum polarization P_r/P_{max} values.[185]

Cycling Stage	No. of Regions	M Fraction	O Fraction	P_r [μC/cm²]	P_{max} [μC/cm²]
Pristine	9	90 % +/-10 %	10 % +/-10 %	5.1	17.9
Woken-up	9	20 % +/-10 %	80 % +/-10 %	12.8	22.7
Fatigued	8	10 % +/-10 %	90 % +/-10 %	12.7	21.1

Compared to conventional (GI)XRD, STEM is limited in sampling capabilities and statistical assessment of phase fractions. However, the target regions are cycled HfO_2 areas beneath top electrode pad of only a few hundred μm width and, thus, spatially too small for beam sizes of common XRD tools without special microspot optics. Consequently, the STEM approach remains competitive and realistic without other easy options at hand. Moreover, the capability of placing acute attention to the interfacial structure situates the chosen STEM approach uniquely over XRD. With the IS results in mind, this might be a very important property to be leveraged.

Several grains of the pristine Gd:HfO_2 sample showed noticeable transitions from the monoclinic or orthorhombic bulk of the grains into (distored) tetragonal symmetry toward the TiN electrode (Fig. 5.23). These transition regions were found to be between half and several unit cells thick and to smoothly relax into the bulk phase of the respective grain as highlighted by the lattice parameter maps (right column) overlaid to the original images (left column). Such interface region were present yet diminished in the samples of woken-up and fatigued state. Certain regions exhibited some relaxations into a tetragonal structure adjacent to the TiN electrodes. Compared to the pristine case, these layer were generally smaller. Their tetragonal symmetry was more subtle and distorted suggesting various stages of relaxation or a confinement at the electrodes. A recent simulation paper hinted toward the generation oxygen vacancies and interstitial O ions (Frenkel defects), during fatigue.[282] Those defects were proposed to give rise to domain wall pinning as the underlying effect of the polarization fatigue. However, an evaluation of the distribution of interatomic distances in the bulk is not able to prove this conclusions as no strong trends (exceeding the uncertainties of the method) were found. More exhaustive STEM details can be found in the supporting information of the corresponding publication.[185]

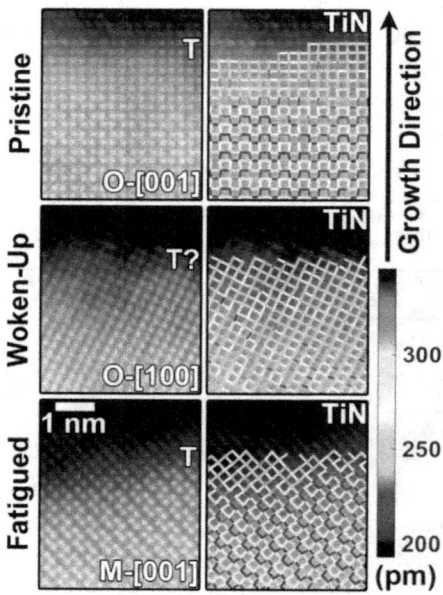

Fig. 5.23: Examples of HAADF-STEM micrographs: The interfacial HfO$_2$ layer undergoes complex changes in interfacial strain and phase presence with field cycling, including some reduction of the presence of the tetragonal phase in the cycled (woken-up and fatigued) samples compared to the pristine capacitor. Lattice parameter maps from these regions further highlight the complex interface environments and assist in visualizing strain and phase presence (colorscale ranges set manually to better visualize details).[185]

Additionally, EELS measurements were performed to assess the potential oxidation of the interfaces or the bulk of the electrodes. Oxygen was present in the 10 nm thick Ti adhesion layer in all three samples—pristine, woken up and fatigued—suggesting a partial but rather uniform oxidation into TiO$_x$. Fig. 5.24 shows the results for the woken-up sample. However, the data indicates that the complete electrodes consisted of a TiO$_x$N$_y$ and also the whole Ti adhesion was layer noticeably oxidized. No conclusive evidence of an interfacial TiO$_x$N$_y$ in the top and bottom electrode could be provided, which is in accord with the findings from the IS. No CPE was necessary to represent this electrode-like layer as would be expected from a high ratio of roughness to thickness of an interfacial layer.

5.6 Structural Evolution During Field Cycling

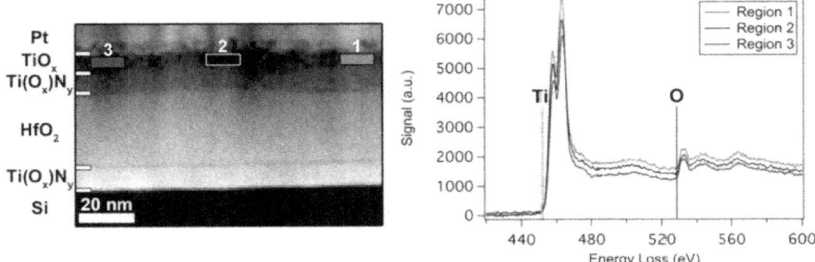

Fig. 5.24: EELS image (left) of the woken-up sample with white tick marks showing layers of the stack, and with colored rectangles 1 – 3 indicating regions in the Ti adhesion layer where the signal from the spectrum image was integrated and plotted (right). The spectra are very similar at each location, indicating relative uniformity of oxygen content. The Ti L-edge and O K-edge onsets are marked, and occur near their known values of 456 eV and 532 eV respectively.[185]

In combination of electrical and STEM results, the structural evolution of the sample can be distinguished into bulk and interface effects. The interface effect most likely does not stem from a TiO_xN_y layer. As EELS results show, R_3, C_3 and R_4 represent the electrodes, an oxidized adhesion layer and the Pt contact pads. The small changes for for R_3, C_3 and R_4, which represent this non-ideal electrode in the equivalent circuit could not be explained by the structural data.

The more important interfacial effect stems from the dielectric interfacial layer found in HfO_2 itself. This (distorted) tetragonal layer diminished during field cycling. However, its conductivity first decreased (both in R_2 and Q_2 contribution), which counteracted and even surpassed the effect of thickness reduction. It can be concluded, that a significantly higher number of charges is present at the beginning. Adjacent to an oxidized electrode, these defects can be assumed to include charged oxygen vacancies. A non-uniform lateral distribution of a defect-rich charged layer at one or both electrodes explains the occurrence of internal bias fields as the origin of the initially constricted P-E hysteresis.[187] A uniform distribution of such charged defects, however, would give rise to an imprint[48] (compare Fig. 2.6 b)). Fig. 5.25 illustrates the effect of these potential scenarios on the macroscopic hysteresis loop. Lomenzo et al. attributed the apparent imprint of a pristine $Si:HfO_2$ to the formation of O-vacancies adjacent to the TaN bottom electrode. Since oxygen vacancies were shown to stabilize the tetragonal compared to the monoclinic bulk phase[172], the inference of the interfacial tetragonal regions being defect-rich and thus, charged regions seems consistent. The aforementioned simulation approach[282] and another TEM study[281] support a redistribution of interfacial oxygen vacancies during field cycling. This scenario is similar to what was termed "surface effect" for perovskite ferroelectrics (see section 5.4 and discussion of Fig. 2.6 g)). As already discussed in section 4.5, a certain templating effect by the TiN electrodes exists. The $Gd:HfO_2$ films are deposited on an already crystalline bottom electrode and also the top electrode is already crystalline before the subsequent annealing step that crystallizes the Gd:HfO$_2$ film. This might result in the distorted tetragonal interface structure which is more compatible to the cubic TiN lattice and relaxes as the strain vanishes toward the bulk.

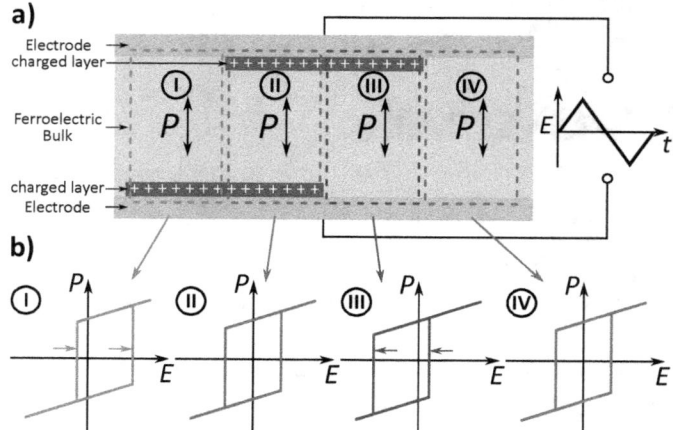

Fig. 5.25: Effect of non-uniform defect-rich, charged layers adjacent to the electrodes: a) Capacitor stack with four different cases of positive charge only at the bottom electrode (I), at both electrodes (II), only at the top electrode (III) and at none of the electrodes (IV). b) Manifestation of the aforementioned scenarios in the macroscopically measures P-E hysteresis. [185]

In the bulk, a structural transformation of monoclinic to orthorhombic hafnia was found. While only 10 % instead of 80 % or 90 % of the grains were orthorhombic in the pristine state, the remanent polarization and maximum polarization were 40 % and 70 % of the respective values after wake-up (1000 cycles). It is likely, that part of the transformation took already place during the first P-E measurement cycle. Moreover, the influence of texture has to be considered when deriving quantitative conclusions. Here, TEM, which depends on a special orientation of grains toward the plane of the prepared lamella is not a suitable approach. Microspot-XRD in Bragg-Brentano geometry might be more suitable, but the limited measurements performed earlier to assess wake-up did not hint at phase changes[172] and thus, no exhaustive texture measurements were performed at that time. The onset of fatigue is marked by the decay in P_r found around 5000 cycles. However, the phase fractions extracted within the statistical limitations of STEM do not indicate a change in phase fractions. This is in accordance to the assumption that during fatigue, domains only get stuck instead of transforming to a non-polar phase. However, the sampling capabilities of the used STEM approach are limited and also hampers credible quantitative conclusions. In section 4.1, first simulation results were summarized that should be taken into account here. The barrier between a tetragonal P4$_2$/nmc and orthorhombic Pca2$_1$ phase was calculated to be 30 meV/f.u. ZrO$_2$.[236] In contrast, from an intermediate state similar to an orthogonalized monoclinic P2$_1$/c phase, several 100 meV are expected.[235] Materlik et al.[61] simulated an effect of −10 meV on the ground state of the orthorhombic Pca2$_1$ for an applied field of 1 MV/cm.

An effect of similar magnitude can be anticipated for the barrier between the non-polar and this polar phase. Additional leverage is expected to result from the redistribution of oxygen vacancies from the interfaces into the bulk. Hoffmann et al.[172] showed an effect of around 25 meV/f.u. for 10 % oxygen vacancies. Overall, the impact of the electric field and oxygen vacancies still seems rather low compared to what seems needed to explain a direct transition from a monoclinic(-like) to the orthorhombic phase. A transition via an intermediate phase (e.g. the tetragonal phase) could take place instead. Barrier calculations of more precisely matched unit cells under applied electric field and with a certain amount of oxygen vacancies included might help solving the current mismatch. These phase changes are not the only cycling-induced changes in the bulk of the film. Increasing leakage current levels indicate an increasing amount of defects in the bulk of the layer. By analogy to the simulation results for $Sr:HfO_2$[282], the creation of defects within the bulk was concluded. While the applied methods do not allow for an explicit proof of their role as pinning centers for ferroelectric domains, a clear correlation of a leakage current increase by orders of magnitude and the onset of fatigue is evident.

5.7 Routes to Improve the Field Cycling Behavior

It has been shown that both fatigue and wake-up are detrimental to the operation of ferroelectric memory devices. Suitable annealing conditions could be one main knob to improve the field cycling behavior.[187] At the moment, in-depth studies are still missing. The fatigue problems of commercial ferroelectric memories based on PZT have been solved by 1) IrO_2 or Ir electrodes[22], which do not pull oxygen out of the ferroelectric film, and 2) an optimization of the stochiometry with respect to the Pb content[276, 277, 59].

The higher the thermal budget, the higher the tendency of oxidizing the TiN electrodes and causing some intermixing of electrode and dielectric material.[278, 279 200] This usually changes the electronic band structure toward a lower effective band offset between electrode Fermi level and the conduction (electrode injection) or valence band (hole injection) of HfO_2.[131, 295] However, for a ferroelectric, the interplay between the formation of a spontaneous polarization upon cooling and the alignment of defects leading to internal bias fields[187] is to be considered in addition to that and is rather complex. No general statement is possible at the current state of understanding.

TiN has been the most commonly used electrode for HfO_2/ZrO_2 based ferroelectrics up to now. This was due to the fact that it has been applied in DRAM capacitors developed at that time[296, 131] and that it is a standard electrode material for so-called "high-k metal gate" transistors[19] in CMOS technology. The role of the electrodes in terms of structural and—of special interest here—chemical effects has been discussed in section 4.5. Although a promoting effect on P_r was reported for TaN, an inert electrode might be a more favorable choice to not impact the film stoichiometry and maybe change it during continuous switching.

Another knob might be a different dopant profile across the film with less doping towards the electrodes to compensate for the lower oxygen content. Oxygen vacancies stabilize the higher symmetry phases. If the FE phase is stable in the middle of the film without oxygen depletion, a trend towards tetragonal/cubic interfaces is to be anticipated. A lower dopant concentration toward the electrodes, could counteract this trend. In general, it would be preferable to avoid oxygen vacancies as much as possible. Even in films without ferroelectric properties, they cause trap levels and leakage currents, that result in a degradation of the dielectric film at some point during device operation.[97] For inert electrodes, a homogeneous dopant concentration profile would be the natural choice according to the current understanding. However, if stress conditions induced by the electrodes are found to strongly alter phase stability, again, the introduction of a gradient should be considered. A recent study used Si and Al doping in one and the same ALD-prepared film.[246] Al and Si were formerly reported to induce different unit cell sizes[235]. The paper attributed the different polarizations achieved by varied layering sequences to a stress profile within the ferroelectric itself, without giving a detailed picture of this profile at that stage of research.

Concluding this section, the most promising routes to follow can be summarized as follows:

- optimized annealing conditions
- inert and oxygen blocking electrodes
- stoichiometry optimization (homogeneous or intentional gradient towards the electrodes).

However, it should be kept in mind that these measures cannot be treated independently from one another.

5.8 Summary

This chapter was devoted to electric field cycling behavior of ferroelectric capacitors based on hafnium oxide thin films. The importance of this behavior for applications such as ferroelectric memories was pointed out, structural origins were elucidated and finally conclusions on how to optimize this behavior were derived.

In sections 5.1 to 5.3, the three effects of 1) wake-up, 2) polarization fatigue and 3) a split-up/merging effect first identified in this work were described. First-order reversal curves helped to identify (complex) bias fields as the reason for constricted hysteresis observed in pristine samples (before wake-up) and after cycling treatment with non-saturation field amplitudes (split-up effect). A modified approach of harmonic analysis (section 3.7) was applied to extract activation energies in the order of 100 meV. Frequency and field dependence of these values as well were discussed in section 5.4. Potential mechanism were derived from

5.8 Summary

studies of similar effects in conventional ferroelectrics as well as hafnia and zirconia specific literature. Moreover, the energy values were critically discussed and the comparability to literature values explained.

To assess the underlying effects, scanning probe methods (section 5.5) as well as a combination of transmission electron microscopy and impedance spectroscopy (section 5.6) were utilized. Scanning probe methods were shown to be prone to perturbing effects. Results that agreed well with the findings of internal bias fields were not clear of artifacts and could thus, not be considered as conclusive evidence of any scenario. Nonetheless, the findings could serve as a guide for design of more suitable experiments and careful crosschecks of the obtained results measured on similar ultrathin films of materials with rather low intrinsic piezoelectric response. Impedance spectroscopy and transmission electron microscopy provided insight into the structural evolution of the $Gd:HfO_2$ films during field cycling. Bulk and interface effects could be identified. The bulk effects consist of phase transformations during wake-up, which explains the increase in P_r and defect generation during fatigue, which was proposed to give rise to domain wall pinning. The interface effect of a ≈ 1 nm thick, defect-rich tetragonal layer adjacent to the TiN electrodes was demonstrated to give rise to the internal bias fields and the consequent pinched hysteresis observed for pristine films.

To improve the field cycling behavior (section 5.7) of capacitors based on these new hafnia/zirconia based ferroelectrics, optimized annealing conditions, inert and oxygen blocking electrodes as well as stoichiometry optimization approaches were proposed. Similarities to routes followed for conventional ferroelectrics exist as they were partially derived by analogy to what literature provided for PZT and co. Additionally, all approaches were discussed in reference to their influence on phase stability as the central topic of the previous chapter 4 to account for material specific challenges.

6 Summary and Conclusion

This thesis consists of two two main pillars. The first main goal was a fundamental understanding of the ferroelectricity in hafnia based thin films. Based on this understanding, a highly application-relevant topic was chosen as second main objective of this thesis: the electric field cycling behavior. This term describes a change of the ferroelectric properties through the course of switching operation. These changes are undesired and thus, elucidating underlying mechanisms is vital to optimizing the material stacks for applications such as ferroelectric memories.

In chapter 4, the basis of how to stabilize and how to impact the just recently discovered ferroelectric behavior in HfO_2 was elaborated. The originally proposed $Pca2_1$ phase was proven as the root of the ferroelectricity. A combination of sophisticated drift-corrected STEM methods and electron diffraction was used. The non-orthorhombic phases could be excluded and the FE phase was distinguished from the remaining non-polar phases by proving its missing centrosymmetry. Together with XRD results and simulations, it was argued that the observed AFE-like hystereses stem from a field-induced phase transition. GIXRD results for high Si content exhibited a tetragonal pattern. Thus, the vanishing P_r for higher Si contents or higher temperatures cannot be attributed to a randomized orientation of polar unit cells as it would be the case for an order-disorder transition. The ferroelectricity is therefore predominantly of displacive nature. Larger dopants such as Sr or Gd are more suitable stabilizers of the FE phase than Si (or Al and hafnia in mixture with ZrO_2), which is a proper choice if AFE-like behavior is intended. The high-symmetry phase induced by small dopants (compared to Hf) is tetragonal, whereas for larger dopants, a (pseudo-)cubic phase is formed.

Besides doping, thickness, anneals and different electrodes were used as experimental knobs to impact the ferroelectric properties of the thin films. A higher thermal budget has a similar effect as a thickness increase and favors the formation of the monoclinic bulk phase. The effect of TiN and TaN electrodes was distinguished into structural (clamping, texture, grain size confinement) and chemical (oxygen vacancies, which alter phase stability) effects. The interaction of these experimental parameters with variables important for theoretical calculations, such as defects and grain size, were discussed. XRD studies revealed a fiber texture in FE hafnia thin films. A studied $Si:HfO_2$ sample did not exhibit a strong preferential orientation. Together with the ferroelectric phase fraction from Rietveld refinement this explains the macroscopically measured P_{sat} value. Finally, it was explained that hafnia as a material cannot be called an incipient ferroelectric in a narrower sense. Hafnia thin films can, however, be modified to achieve a Curie temperature slightly below 0 K to mimic the behavior of incipient ferroelectrics. These films, that are subjected to certain specific boundary conditions to suppress the bulk properties, could be called incipiently ferroelectric in a wider sense of the word.

In the beginning of chapter 5, the term "electric field cycling behavior" was introduced as generic name for three effects: wake-up, fatigue and a split-up phenomenon of transient current peaks to be explained later on. "Wake-up" refers to an often observed hysteresis depinching and P_r increase during the initial switching cycles. Fatigue describes the decrease of P_r at higher numbers of switching cycles, which is due to domains that cannot participate in the switching process anymore. A novel split-up/merging phenomenon of transient current peaks was found. Field cycling with non-saturating amplitude(s) could split up a single peak into multiple peaks. Subsequent field cycling with saturating amplitude could merge these peaks back into one single peak again and re-establish the situation before the split-up. All three effects were described and activation energies were derived using a modified approach of harmonic analysis. The derived energies were discussed with respect to other known energies of e.g. ion migration and phase transitions. Pros and cons of harmonic analysis and other methods to derive activation energies were contrasted, but no solely true and best method exists. A major problem is the temporal interference of different phenomena during measurement. Next, FORC measurements were used to reveal that internal bias fields are the root of hysteresis pinching in both wake-up and the split-up scenario. For the wake-up, a combination of STEM and impedance spectroscopy could show that these bias fields stem from defect rich charged layers adjacent to the TiN electrodes. These layers were found to be inhomogenously distributed and consisted of tetragonally distorted HfO_2. Moreover, the FE orthorhombic fraction of the bulk phase was higher after wake-up than in the pristine case. During polarization fatigue, no bias fields or phase changes were observed. Simulations and a correlation with increasing leakage currents made a defect generation in the bulk of the ferroelectric thin films a very likely origin of fatigue. These defects were proposed to give rise to domain wall pinning. Further, the detrimental role of internal bias fields for low-voltage operation of ferroelectric memories was explained. Promising routes to optimize both wake-up and fatigue were derived and potential compromises to be made with respect to phase stability were pointed out. Finally, an SPM study, which was also supposed to shed light into the field-cycling effects on nanoscale, was only able to provide hints at the same effects as the FORC, TEM and impedance measurements. However, it unveiled the critical role of perturbing effects. Suitable experiments to cross-check for artifacts were listed and performed within this work. Caution is advised when interpreting similar PFM experiments.

Combined, the results of presented in these two chapters help advancing the fundamental material understanding on the one hand. On the other hand, they present a significant step in paving the way for exploiting ferroelectric properties of hafnia and zirconia in applications such as ferroelectric memories. The potential of these materials has already been acknowledged by the International Technology Roadmap for Semiconductors.[106] While in the field of ferroelectric memories mainly sophisticated engineering efforts on real memory stacks and structures are needed, fundamental research on pyroelectric, piezoelectric and optical properties of the $Pca2_1$ phase and corresponding applications remain still unexploited fields of research.

Bibliography

[1] T. S. Böscke, J. Müller, D. Bräuhaus, U. Schröder, and U. Böttger. Ferroelectricity in hafnium oxide thin films. *Applied Physics Letters*, 99(10):102903, 2011.

[2] G. W. Burr, B. N. Kurdi, J. C. Scott, C. H. Lam, K. Gopalakrishnan, and R. S. Shenoy. Overview of candidate device technologies for storage-class memory. *IBM Journal of Research and Development*, 52(4.5):449–464, 2008.

[3] R. F. Freitas and W. W. Wilcke. Storage-class memory: The next storage system technology. *IBM Journal of Research and Development*, 52(4.5):439–447, 2008.

[4] J. Edmondson, W. Anderson, J. Gray, J. P. Loyall, K. Schmid, and J. White. Next-generation mobile computing. *IEEE Software*, 31(2):44–47, 2014.

[5] G. S. Kearns. Countering Mobile Device Threats: A Mobile Device Security Model. *Journal of Forensic & Investigative Accounting*, 8(1), 2016.

[6] J. Gubbi, R. Buyya, S. Marusic, and M. Palaniswami. Internet of Things (IoT): A vision, architectural elements, and future directions. *Future Generation Computer Systems*, 29(7):1645–1660, 2013.

[7] D. Miorandi, S. Sicari, F. De Pellegrini, and I. Chlamtac. Internet of things: Vision, applications and research challenges. *Ad Hoc Networks*, 10(7):1497–1516, 2012.

[8] M. Aazam, I. Khan, A. A. Alsaffar, and E.-N. Huh. Cloud of Things: Integrating Internet of Things and cloud computing and the issues involved. In *11th International Bhurban Conference on Applied Sciences and Technology (IBCAST), 2014*, pages 414–419. IEEE, 2014.

[9] R. Want, B. N. Schilit, and S. Jenson. Enabling the Internet of Things. *Computer*, 48(1):28–35, 2015.

[10] G. E. Moore. Cramming More Components Onto Integrated Circuits. *Proceedings of the IEEE*, 86(1):82–85, 1998.

[11] S. Joseph, V. Namboodiri, and V. C. Dev. A MarketDriven Framework Towards Environmentally Sustainable Mobile Computing. *ACM SIGMETRICS Performance Evaluation Review*, 42(3):46–48, 2014.

[12] G. I. Meijer. Cooling Energy-Hungry Data Centers. *Science*, 328(5976):318–319, 2010.

[13] Z. Li and S. G. Kandlikar. Current Status and Future Trends in Data-Center Cooling Technologies. *Heat Transfer Engineering*, 36(6):523–538, 2015.

[14] Texas Instruments. FRAM – New Generation of Non-Volatile Memory Bulletin, White Paper, http://www.ti.com/lit/ml/szzt014a/szzt014a.pdf. 2009.

[15] S. R. Summerfelt, T. S. Moise, K. R. Udayakumar, K. Boku, K. Remack, J. Rodriguez, J. Gertas, H. McAdams, S. Madan, J. Eliason, and others. High-density 8mb 1t-1c ferroelectric random access memory embedded within a low-power 130nm logic process. In *16th IEEE International Symposium on Applications of Ferroelectrics (ISAF), 2007*, pages 9–10. IEEE, 2007.

[16] J. Rodriguez, K. Remack, J. Gertas, L. Wang, C. Zhou, K. Boku, J. Rodriguez-Latorre, K. R. Udayakumar, S. Summerfelt, T. Moise, D. Kim, J. Groat, J. Eliason, M. Depner, and F. Chu. Reliability of Ferroelectric Random Access memory embedded within 130nm CMOS. pages 750–758. IEEE, 2010.

[17] Texas Instruments. Texas Instruments Delivers First Chip Made or Advanced 90nm Process, press release, http://newscenter.ti.com/news-releases?item=126318. January 2003.

[18] Y. K. Hong, D. J. Jung, S. K. Kang, H. S. Kim, J. Y. Jung, H. K. Koh, J. H. Park, D. Y. Choi, S. E. Kim, W. S. Ann, and others. 130 nm-technology, 0.25 μm^2, 1T1C FRAM Cell for SoC (System-on-a-Chip)-friendly Applications. In *IEEE Symposium on VLSI Technology, 2007*, pages 230–231. IEEE, 2007.

[19] M. Bohr, R. Chau, T. Ghani, and K. Mistry. The High-k Solution. *IEEE Spectrum*, 44(10):29–35, 2007.

[20] Analogies and Differences between Ferroelectrics and Ferromagnets. In K. Rabe, C. H. Ahn, J.-M. Triscone, and Nicola A. Spaldin, editors, *Physics of Ferroelectrics: A Modern Perspective*, number 105 in Topics Appl. Physics, pages 175–218 Springer-Verlag Berlin Heidelberg, 2007.

[21] U. Böttger. Dielectric Properties of Polar Oxides. In R. Waser, U. Böttger, and S. Tiedke, editors, *Polar Oxides: Properties, Characterization, and Imaging*. Wiley-VCH Verlag GmbH & Co. KGaA, Weinheim, 2006.

[22] J. F. Scott. *Ferroelectric Memories*, volume 3 of *Springer Series in Advanced Microelectronics*. Springer Berlin Heidelberg, Berlin, Heidelberg, 2000.

[23] T. Mikolajick, S. Müller, T. Schenk, E. Yurchuk, S. Slesazeck, U. Schröder, S. Flachowsky, R. van Bentum, S. Kolodinski, P. Polakowski, and J. Müller. Doped Hafnium Oxide – An Enabler for Ferroelectric Field Effect Transistors. *Advances in Science and Technology*, 95:136–145, 2014.

[24] P. Groth. Ueber Beziehungen zwischen Krystallform und chemische Constitution bei einigen organischen Verbindungen. *Annalen der Physik*, 217(9):31–43, 1870.

[25] V. M. Goldschmidt. Crystal structure and chemical constitution. *Transactions of the Faraday Society*, 25:253–283, 1929.

[26] N. Inoue, T. Takeuchi, and Y. Hayashi. Sputtering process design of PZT capacitors for stable FeRAM operation. In *Technical Digest., International Electron Devices Meeting (IEDM), 1998*, pages 819–822. IEEE, 1998.

[27] E. Y. Tsymbal, E. R. A. Dagotto, C.-B. Eom, and R. Ramesh. *Multifunctional Oxide Heterostructures*. OUP Oxford, 2012.

[28] A Landau Primer for Ferroelectrics. In K. Rabe, C. H. Ahn, J.-M. Triscone, Premi Chandra, and Peter B. Littlewood, editors, *Physics of Ferroelectrics: A Modern Perspective*, number 105 in Topics Appl. Physics, pages 69–116. Springer-Verlag Berlin Heidelberg, 2007.

[29] A. F. Devonshire. XCVI. Theory of barium titanate: Part I. *The London, Edinburgh, and Dublin Philosophical Magazine and Journal of Science*, 40(309):1040–1063, 1949.

[30] A. F. Devonshire. CIX. Theory of barium titanate: Part II. *The London, Edinburgh, and Dublin Philosophical Magazine and Journal of Science*, 42(333):1065–1079, 1951.

[31] M. E. Lines and A. M. Glass. *Principles and applications of ferroelectrics and related materials*. The International series of monographs on physics. Clarendon Press, Oxford, England, 1977.

[32] F. Preisach. Ü die magnetische Nachwirkung. *Zeitschrift für Physik*, 94(5-6):277–302, 1935.

[33] A. Stancu, D. Ricinschi, L. Mitoseriu, P. Postolache, and M. Okuyama. First-order reversal curves diagrams for the characterization of ferroelectric switching. *Applied Physics Letters*, 83(18):3767, 2003.

[34] I. D. Mayergoyz. Scalar Preisach models of hysteresis. In G. Bertotti and I. D. Mayergoyz, editors, *The Science of Hysteresis: Mathematical modeling and applications*, volume 1. Academic, Amsterdam; Boston, 1 edition, 2006.

[35] T. Olsen, U. Schröder, S. Müller, A. Krause, D. Martin, A. Singh, J. Müller, M. Geidel, and T. Mikolajick. Co-sputtering yttrium into hafnium oxide thin films to produce ferroelectric properties. *Applied Physics Letters*, 101(8):082905, 2012.

[36] D. Zhou, J. Xu, Q. Li, Y. Guan, F. Cao, X. Dong, J. Müller, T. Schenk, and U. Schröder. Wake-up effects in Si-doped hafnium oxide ferroelectric thin films. *Applied Physics Letters*, 103(19):192904, 2013.

[37] U. Schroeder, E. Yurchuk, J. Müller, D. Martin, T. Schenk, P. Polakowski, C. Adelmann, M. I. Popovici, S. V. Kalinin, and T. Mikolajick. Impact of different dopants on the switching properties of ferroelectric hafniumoxide. *Japanese Journal of Applied Physics*, 53(8S1):08LE02, 2014.

[38] M. I. Morozov and D. Damjanovic. Hardening-softening transition in Fe-doped Pb(Zr,Ti)O$_3$ ceramics and evolution of the third harmonic of the polarization response. *Journal of Applied Physics*, 104(3):034107, 2008.

[39] Julia Glaum, Yuri A. Genenko, Hans Kungl, Ljubomira Ana Schmitt, and Torsten Granzow. De-aging of Fe-doped lead-zirconate-titanate ceramics by electric field cycling: 180°- vs. non-180°domain wall processes. *Journal of Applied Physics*, 112(3):034103, 2012.

[40] T. M. Kamel and G. de With. Poling of hard ferroelectric PZT ceramics. *Journal of the European Ceramic Society*, 28(9):1827–1838, 2008.

[41] K. Carl and K. H. Hardtl. Electrical after-effects in Pb(Ti, Zr)O$_3$ ceramics. *Ferroelectrics*, 17(1):473–486, 1977.

[42] V. Ya. Shur, I. S. Baturin, and E. L. Rumyantsev. Analysis of the Switching Data in Inhomogeneous Ferroelectrics. *Ferroelectrics*, 349(1):163–170, 2007.

[43] G. Arlt and U. Robels. Aging and fatigue in bulk ferroelectric perovskite ceramics. *Integrated Ferroelectrics*, 3(4):343–349, 1993.

[44] Q. Jiang, W. Cao, and L. E. Cross. Electric Fatigue in Lead Zirconate Titanate Ceramics. *Journal of the American Ceramic Society*, 77(1):211–215, 1994.

[45] E. Sawaguchi and M. L. Charters. Aging and the Double Hysteresis Loop of Pb$_\lambda$Ca$_{1-\lambda}$TiO$_3$ Ceramics. *Journal of the American Ceramic Society* 42(4):157–164, 1959.

[46] D. Lin, K. W. Kwok, and H. L. W. Chan. Double hysteresis loop in Cu-doped K$_{0.5}$Na$_{0.5}$NbO$_3$ lead-free piezoelectric ceramics. *Applied Physics Letters*, 90(23):232903, 2007.

[47] G. L. Yuan, Y. Yang, and S. W. Or. Aging-induced double ferroelectric hysteresis loops in BiFeO$_3$ multiferroic ceramic. *Applied Physics Letters*, 91(12):122907, 2007.

[48] T. Schenk, E. Yurchuk, S. Mueller, U. Schroeder, S. Starschich, U. Böttger, and T. Mikolajick. About the deformation of ferroelectric hystereses. *Applied Physics Reviews*, 1(4):041103, 2014.

[49] J. F. Scott. Ferroelectrics go bananas. *Journal of Physics: Condensed Matter*, 20(2):021001, 2008.

[50] R. Meyer, R. Waser, K. Prume, T. Schmitz, and S. Tiedke. Dynamic leakage current compensation in ferroelectric thin-film capacitor structures. *Applied Physics Letters*, 86(14):142907, 2005.

[51] T. Schenk, U. Schroeder, and T. Mikolajick. Correspondence - Dynamic leakage current compensation revisited. *IEEE Transactions on Ultrasonics, Ferroelectrics, and Frequency Control*, 62(3):596–599, 2015.

[52] G. Le Rhun, R. Bouregba, and G. Poullain. Polarization loop deformations of an oxygen deficient $Pb(Zr_{0.25},Ti_{0.75})O_3$ ferroelectric thin film. *Journal of Applied Physics*, 96(10):5712, 2004.

[53] D. Dimos, W. L. Warren, M. B. Sinclair, B. A. Tuttle, and R. W. Schwartz. Photoinduced hysteresis changes and optical storage in $(Pb,La)(Zr,Ti)O_3$ thin films and ceramics. *Journal of Applied Physics*, 76(7):4305, 1994.

[54] U. Robels, J. H. Calderwood, and G. Arlt. Shift and deformation of the hysteresis curve of ferroelectrics by defects: An electrostatic model. *Journal of Applied Physics*, 77(8):4002, 1995.

[55] M. Stengel and N. A. Spaldin. Origin of the dielectric dead layer in nanoscale capacitors. *Nature*, 443(7112):679–682, 2006.

[56] A. K. Tagantsev, M. Landivar, E. Colla, and N. Setter. Identification of passive layer in ferroelectric thin films from their switching parameters. *Journal of Applied Physics*, 78(4):2623, 1995.

[57] T. P. Ma and J.-P. Han. Why is nonvolatile ferroelectric memory field-effect transistor still elusive? *IEEE Electron Device Letters*, 23(7):386–388, 2002.

[58] A. K. Tagantsev, I. Stolichnov, E. L. Colla, and N. Setter. Polarization fatigue in ferroelectric films: Basic experimental findings, phenomenological scenarios, and microscopic features. *Journal of Applied Physics*, 90(3):1387, 2001.

[59] X. J. Lou. Polarization fatigue in ferroelectric thin films and related materials. *Journal of Applied Physics*, 105(2):024101, 2009.

[60] C. Kittel. Theory of antiferroelectric crystals. *Physical Review*, 82(5):729, 1951.

[61] R. Materlik, C. Künneth, and A. Kersch. The origin of ferroelectricity in $Hf_{1-x}Zr_xO_2$: A computational investigation and a surface energy model. *Journal of Applied Physics*, 117(13):134109, 2015.

[62] Z. Feng and X. Ren. Aging effect and large recoverable electrostrain in Mn-doped $KNbO_3$-based ferroelectrics. *Applied Physics Letters*, 91(3):032904, 2007.

[63] P. V. Lambeck and G. H. Jonker. The nature of domain stabilization in ferroelectric perovskites. *Journal of Physics and Chemistry of Solids*, 47(5):453–461, 1986.

[64] C. F. Pulvari. Ferrielectricity. *Physical Review*, 120(5):1670, 1960.

[65] B. Peng, H. Fan, and Q. Zhang. A Giant Electrocaloric Effect in Nanoscale Antiferroelectric and Ferroelectric Phases Coexisting in a Relaxor $Pb_{0.8}Ba_{0.2}ZrO_3$ Thin Film at Room Temperature. *Advanced Functional Materials*, 23(23):2987–2992, 2013.

[66] B. Peng, H. Fan, and Q. Zhang. High Tunability in (111)-Oriented Relaxor $Pb_{0.8}Ba_{0.2}ZrO_3$ Thin Film with Antiferroelectric and Ferroelectric Two-Phase Coexistence. *Journal of the American Ceramic Society*, 96(6):1852–1856, 2013.

[67] D. Viehland. Transmission electron microscopy study of high-Zr-content lead zirconate titanate. *Physical Review B*, 52(2):778–791, 1995.

[68] J. E. Daniels, W. Jo, J. Rödel, and J. L. Jones. Electric-field-induced phase transformation at a lead-free morphotropic phase boundary: Case study in a 93%$(Bi_{0.5}Na_{0.5})TiO_3$-7%$BaTiO_3$ piezoelectric ceramic. *Applied Physics Letters*, 95(3):032904, 2009.

[69] X. Tan, C. Ma, J. Frederick, S. Beckman, and K. G. Webber. The Antiferroelectric ↔ Ferroelectric Phase Transition in Lead-Containing and Lead-Free Perovskite Ceramics. *Journal of the American Ceramic Society*, 94(12):4091–4107, 2011.

[70] T. Schenk, U. Schroeder, M. Pešić, M. Popovici, Y. V. Pershin, and T. Mikolajick. Electric Field Cycling Behavior of Ferroelectric Hafnium Oxide. *ACS Applied Materials & Interfaces*, 6(22):19744–19751, 2014.

[71] N. Menou, Ch. Muller, I. S. Baturin, V. Ya. Shur, and J.-L. Hodeau. Polarization fatigue in $PbZr_{0.45}Ti_{0.55}O_3$-based capacitors studied from high resolution synchrotron x-ray diffraction. *Journal of Applied Physics*, 97(6):064_08, 2005.

[72] M. Kohli, P. Muralt, and N. Setter. Removal of 90° domain pinning in (100) $Pb(Zr_{0.15}Ti_{0.85})O_3$ thin films by pulsed operation. *Applied Physics Letters*, 72(24):3217, 1998.

[73] G. M. Guro, I. I. Ivanchik, and N. F. Kovtonyuk. Semiconducting Properties of Ferroelectrics. *Sov. J. Exp. Theor. Phys. Lett. (Pis'ma Zh. Eksp. Teor. Fiz.)*, 5(1):5–8, 1967.

[74] I. P. Batra and B. D. Silverman. Thermodynamic stability of thin ferroelectric films. *Solid State Communications*, 11(1):291–294, 1972.

[75] P. Ghosez and J. Junquera. First-principles modeling of ferroelectric oxide nanostructures. In *Handbook of theoretical and computational nanotechnology*, pages 623–728. American Scientific Publisher, Stevenson Ranch, California, United States, 2006.

[76] L. Cima, E. Laboure, and P. Muralt. Characterization and model of ferroelectrics based on experimental Preisach density. *Review of Scientific Instruments*, 73(10):3546, 2002.

[77] L. Mitoseriu, L. Stoleriu, A. Stancu, C. Galassi, and V. Buscaglia. First order reversal curves diagrams for describing ferroelectric switching characteristics. *Processing and Application of Ceramics*, 3(1-2):3–7, 2009.

[78] V. Ya. Shur, E. V. Nikolaeva, E. I. Shishkin, I. S. Baturin, A. G. Shur, T. Utschig, T. Schlegel, and D. C. Lupascu. Deaging in $Gd_2(MoO_4)_3$ by cyclic motion of a single planar domain wall. *Journal of Applied Physics*, 98(7):074106, 2005.

[79] Marin Alexe, Alexei Gruverman, Phaedon Avouris, Klaus von Klitzing, Hiroyuki Sakaki, and Roland Wiesendanger, editors. *Nanoscale Characterisation of Ferroelectric Materials*. NanoScience and Technology. Springer Berlin Heidelberg, Berlin, Heidelberg, 2004.

[80] A. Gruverman, D. Wu, and J. F. Scott. Piezoresponse Force Microscopy Studies of Switching Behavior of Ferroelectric Capacitors on a 100-ns Time Scale. *Physical Review Letters*, 100(9), 2008.

[81] E. Soergel. Piezoresponse force microscopy (PFM). *Journal of Physics D: Applied Physics*, 44(46):464003, 2011.

[82] A. Roelofs, U. Böttger, R. Waser, F. Schlaphof, S. Trogisch, and L. M. Eng. Differentiating 180°and 90°switching of ferroelectric domains with three-dimensional piezoresponse force microscopy. *Applied Physics Letters*, 77(21):3444, 2000.

[83] L. F. Schloss, P. C. McIntyre, B. C. Hendrix, S. M. Bilodeau, J. F. Roeder, and S. R. Gilbert. Oxygen tracer studies of ferroelectric fatigue in $Pb(Zr,Ti)O_3$ thin films. *Applied Physics Letters*, 81(17):3218, 2002.

[84] C. U. Pinnow, I. Kasko, N. Nagel, T. Mikolajick, C. Dehm, F. Jahnel, M. Seibt, U. Geyer, and K. Samwer. Oxygen tracer diffusion in IrO_2 barrier films. *Journal of Applied Physics*, 91(3):1707, 2002.

[85] R.-V. Wang and P. C. McIntyre. ^{18}O tracer diffusion in $Pb(Zr,Ti)O_3$ thin films: A probe of local oxygen vacancy concentration. *Journal of Applied Physics*, 97(2):023508, 2005.

[86] X. Ren. Large electric-field-induced strain in ferroelectric crystals by point-defect-mediated reversible domain switching. *Nature Materials*, 3(2):91–94, 2004.

[87] W. L. Warren, D. Dimos, G. E. Pike, K. Vanheusden, and R. Ramesh. Alignment of defect dipoles in polycrystalline ferroelectrics. *Applied Physics Letters*, 67(12):1689, 1995.

[88] R. V. Vedrinskii, V. L. Kraizman, A. A. Novakovich, Ph. V. Demekhin, and S. V. Urazhdin. Pre-edge fine structure of the 3d atom K x-ray absorption spectra and quantitative atomic structure determinations for ferroelectric perovskite structure crystals. *Journal of Physics: Condensed Matter*, 10(42):9561–9580, 1998.

[89] N. Ikeda, H. Ohsumi, K. Ohwada, K. Ishii, Toshiya Inami, K. Kakurai, Y. Murakami, K. Yoshii, S. Mori, Y. Horibe, and H. Kitô. Ferroelectricity from iron valence ordering in the charge-frustrated system $LuFe_2O_4$. *Nature*, 436(7054):1136–1138, August 2005.

[90] M. H. Frey and D. A. Payne. Grain-size effect on structure and phase transformations for barium titanate. *Physical Review B*, 54(5):3158–3168, 1996.

[91] M. D. Fontana, G. Metrat, J. L. Servoin, and F. Gervais. Infrared spectroscopy in $KNbO_3$ through the successive ferroelectric phase transitions. *Journal of Physics C: Solid State Physics*, 17(3):483–514, 1984.

[92] G. F. Burr. Storage Class Memory (Short Course), IEEE International Interconnect Technology Conference, http://geoffreyburr.org/papers/Burr_IITC2010_slides.pdf, June 2010.

[93] T. Schenk, M. Hoffmann, C. Richter, M. Pešić, S. Mueller, S. Slesazeck, U. Schroeder, T. Mikolajick, D. Pohl, J. Müller, P. Polakowski, R. Materlik, A. Kersch, X. Sang, E. D. Grimley, and J. M. LeBeau. Doped Hafnium Oxide for Ferroelectric Memories, Materials Research Society, Fall Meeting, Boston, MA, http://dx.doi.org/10.13140/RG.2.1.3912.4567, 2015.

[94] Intel Corporation and Micron Technology, Inc. Intel and Micron Produce Breakthrough Memory Technology, press release, http://newsroom.intel.com/docs/DOC-6713. July 2015.

[95] D. Hession, N. Mc Kelvey, and K. Curran. Storage Class Memory. *International Journal of E-Business Development*, 4(1):30–33, 2013.

[96] M. Darcy and M. Ross. IBM Details Next Generation of Storage Innovation – IBM Outlines Intelligent Storage Devices and Storage-Class Memory. Technical report, http://www-03.ibm.com/press/us/en/pressrelease/20209.wss, 2006.

[97] Rainer Waser. *Nanoelectronics and Information Technology*. Wiley-VCH Verlag GmbH & Co. KGaA, Weinheim, 2012.

[98] H. Schroeder, V. V. Zhirnov, R. K. Cavin, and R. Waser. Voltage-time dilemma of pure electronic mechanisms in resistive switching memory cells. *Journal of Applied Physics*, 107(5):054517, 2010.

[99] G. R. Fox, M. Vaudin, T. Maitland, and T. Moise. Thin Films Texture and Scaling Effects in Ferroelectric Random Access Memory, Gordon Research Conference (Ceramics, Solid State Studies), https://www.grc.org/programs.aspx?id=7983, August 2006.

[100] T. Moise, S. R. Summerfelt, K. R. Udayakumar, F. G. Celii, G. Shinn, K. Boku, K. Remack, A. Haider, G. Albrecht, J. S. Martin, J. Rodriguez, J. Certas, B. Khan, S. Aggarwal, N. Schauer, H. McAdams, S. Madan, A. McKerrow, J. Eliason, J. Groat, R. Bailey, G. R. Fox, D. Kim, P. Staubs, M. Depner, and J. Walbert. High Density 8 Mb 1T-1C FRAM Embedded Within a Low-Power 130 nm Logic Process, 15th IEEE International Symposium on Applications of Ferroelectrics, Sunset Beach, NC, United States, August 2006.

[101] H. Ishiwara. Ferroelectric Random Access Memories. *Journal of Nanoscience and Nanotechnology*, 12(10):7619–7627, 2012.

[102] D. A. Buck. Ferroelectrics for digital information storage and switching, master thesis, MIT Digital Computer Laboratory. 1952.

[103] International Technology Roadmap for Semiconductors. Process Integration, Devices, and Structures. Technical report, http://www.itrs2.net/itrs-reports.html, 2013.

[104] J. Hutchby and M. Garner. Assessment of the Potential & Maturity of Selected Emerging Research Memory Technologies Workshop & ERD/ERM Working Group Meeting, April 2010.

[105] International Technology Roadmap for Semiconductors. Emerging Research Devices. Technical report, http://www.itrs2.net/itrs-reports.html, 2011.

[106] International Technology Roadmap for Semiconductors. Emerging Research Devices. Technical report, http://www.itrs2.net/itrs-reports.html, 2013.

[107] J. Müller, P. Polakowski, S. Mueller, and T. Mikolajick. Ferroelectric Hafnium Oxide Based Materials and Devices: Assessment of Current Status and Future Prospects. *ECS Journal of Solid State Science and Technology*, 4(5):N30–N35, 2015.

[108] T.-S. Jung, Y.-J. Choi, K.-D. Suh, B.-H. Suh, J.-K. Kim, Y.-H. Lim, Y.-N. Koh, J.-W. Park, K.-J. Lee, J.-H. Park, K.-T. Park, J.-R. Kim, J.-H. Yi, and H.-K. Lim. A 117-mm^2 3.3-V only 128-Mb multilevel NAND flash memory for mass storage applications. *IEEE Journal of Solid-State Circuits*, 31(11):1575–1583, 1996.

[109] Y. Yamauchi, Y. Kamakura, and T. Matsuoka. Scalable Virtual-Ground Multilevel-Cell Floating-Gate Flash Memory. *IEEE Transactions on Electron Devices*, 60(8):2518–2524, 2013.

[110] Cactus Technologies. SLC vs MLC NAND and The Impact of Technology - White Paper CTWP010. Technical report, http://www.cactus-tech.com/en/resources/white-papers, 2013.

[111] Hafnium(IV) oxide, Wikipedia – The Free Encyclopedia, https://en.wikipedia.org/wiki/Hafnium(IV)_oxide (2015-12-03).

[112] E. A. Mojaki. *Study of the Zirconia-Hafnia System and Particularly its Behaviour at High Temperatures and High Pressure*. PhD thesis, Faculty of Engineering and the Built Environment, University of the Witwatersrand, 2005.

[113] S. Lee, S. Bang, S. Jeon, S. Kwon, W. Jeong, S. Kim, and H. Jeon. Characteristics of Hafnium-Zirconium-Oxide Film Treated by Remote Plasma Nitridation. *Journal of The Electrochemical Society*, 155(7):H516, 2008.

[114] J. C. Hackley and T. Gougousi. Properties of atomic layer deposited HfO_2 thin films. *Thin Solid Films*, 517(24):6576–6583, 2009.

[115] J. Aarik, A. Aidla, H. Mändar, Väino Sammelselg, and Teet Uustare. Texture development in nanocrystalline hafnium dioxide thin films grown by atomic layer deposition. *Journal of Crystal Growth*, 220(1-2):105–113, 2000.

[116] V. V. Kharton, E. N. Naumovich, and A. A. Vecher. Research on the electrochemistry of oxygen ion conductors in the former Soviet Union. I. ZrO_2-based ceramic materials. *Journal of Solid State Electrochemistry*, 3(2):61–81, 1999.

[117] V. V. Kharton, A. A. Yaremchenko, E. N. Naumovich, and F. M. B. Marques. Research on the electrochemistry of oxygen ion conductors in the former Soviet Union III. HfO_2-, CeO_2- and ThO_2-based oxides. *Journal of Solid State Electrochemistry*, 4(5):243–266, 2000.

[118] R. C. Garvie, R. H. Hannink, and R. T. Pascoe. Ceramic steel? *Nature*, 258(5537):703–704, 1975.

[119] Carl E. Curtis. Development of Zirconia Resistant to Thermal Shock. *Journal of the American Ceramic Society*, 30(6):180–196, 1947.

[120] C. E. Curtis, L. M. Doney, and J. R. Johnson. Some Properties of Hafnium Oxide, Hafnium Silicate, Calcium Hafnate, and Hafnium Carbide. *Journal of the American Ceramic Society*, 37(10):458–465, 1954.

[121] R. Ruh, H. J. Garrett, R. F. Domagala, and NMf Tallan The System Zirconia-Hafnia. *Journal of the American Ceramic Society*, 51(1):23–28, 1968.

[122] C. Wang, M. Zinkevich, and F. Aldinger. The Zirconia-Hafnia System DTA Measurements and Thermodynamic Calculations. *Journal of the American Ceramic Society*, 89(12):3751–3758, 2006.

[123] E. H. Kisi and C.J. Howard. Crystal Structures of Zirconia Phases and their Inter-Relation. *Key Engineering Materials*, 153-154:1–36, 1998.

[124] H. Arashi. Pressure-Induced Phase Transformation of HfO_2. *Journal of the American Ceramic Society*, 75(4):844–847, 1992.

[125] M. A. Laguna-Bercero. Recent advances in high temperature electrolysis using solid oxide fuel cells: A review. *Journal of Power Sources*, 203:4–16, 2012.

[126] M. C. Tucker. Progress in metal-supported solid oxide fuel cells: A review. *Journal of Power Sources*, 195(15):4570–4582, 2010.

[127] L. Goux, P. Czarnecki, Y. Y. Chen, L. Pantisano, X. P. Wang, R. Degraeve, B. Govoreanu, M. Jurczak, D. J. Wouters, and L. Altimime. Evidences of oxygen-mediated resistive-switching mechanism in TiN \HfO$_2$\Pt cells. *Applied Physics Letters*, 97(24):243509, 2010.

[128] J. Robertson. High dielectric constant oxides. *The European Physical Journal Applied Physics*, 28(3):265–291, 2004.

[129] H. J. Cho, Y. D. Kim, D. S. Park, E. Lee, C. H. Park, J. S. Jang, K. B. Lee, H. W. Kim, Y. J. Ki, I. K. Han, and Y. W. Song. New TIT capacitor with ZrO$_2$/Al$_2$O$_3$/ZrO$_2$ dielectrics for 60nm and below DRAMs. *Solid-State Electronics*, 51(11-12):1529–1533, 2007.

[130] M. Pešić, S. Slesazeck, T. Schenk, U. Schroeder, and T. Mikolajick. Impact of charge trapping on the ferroelectric switching behavior of doped HfO$_2$: Trapping influence on the ferroelectric switching. *physica status solidi (a)*, pages 270–273, 2015.

[131] M. Pešić, S. Knebel, K. Cho, C. Jung, J. Chang, H. Lim, N. Kolomiiets, V. V. Afanas'ev, T. Mikolajick, and U. Schroeder. Conduction barrier offset engineering for DRAM capacitor scaling. *Solid-State Electronics*, 2015.

[132] J. Müller, U. Schröder, T. S. Böscke, I. Müller, U. Böttger, L. Wilde, J. Sundqvist, M. Lemberger, P. Kücher, T. Mikolajick, and L. Frey. Ferroelectricity in yttrium-doped hafnium oxide. *Journal of Applied Physics*, 110(11):114113, 2011.

[133] X. Zhao and D. Vanderbilt. First-principles study of structural, vibrational, and lattice dielectric properties of hafnium oxide. *Physical Review B*, 65(23), 2002.

[134] Y. Watanabe, H. Ota, S. Migita, Y. Kamimuta, K. Iwamoto, M. Takahashi, A. Ogawa, H. Ito, T. Nabatame, and A. Toriumi. Achievement of Higher-k and High-ϕ in Phase Controlled HfO$_2$ Film Using Post Gate Electrode Deposition Annealing. volume 11, pages 35–45. ECS, 2007.

[135] C. Adelmann, H. Tielens, D. Dewulf, A. Hardy, D. Pierreux, J. Swerts, E. Rosseel, X. Shi, M. K. Van Bael, J. A. Kittl, and S. Van Elshocht. Atomic Layer Deposition of Gd-Doped HfO$_2$ Thin Films. *Journal of The Electrochemical Society*, 157(4):G105, 2010.

[136] A. Toriumi, K. Kita, K. Tomida, and Y. Yamamoto. Doped HfO$_2$ for Higher-k Dielectrics. volume 1, pages 185–197. ECS, 2006.

[137] E. H. Kisi, C. J. Howard, and R. J. Hill. Crystal structure of orthorhombic zirconia in partially stabilized zirconia. *Journal of the American Ceramic Society*, 72(9):1757–1760, 1989.

[138] S. Mueller, J. Mueller, A. Singh, S. Riedel, J. Sundqvist, U. Schroeder, and T. Mikolajick. Incipient Ferroelectricity in Al-Doped HfO$_2$ Thin Films. *Advanced Functional Materials*, 22(11):2412–2417, 2012.

[139] S. Mueller, C. Adelmann, A. Singh, S. Van Elshocht, U. Schroeder, and T. Mikolajick. Ferroelectricity in Gd-Doped HfO$_2$ Thin Films. *ECS Journal of Solid State Science and Technology*, 1(6):N123–N126, 2012.

[140] J. Müller, T. S. Böscke, S. Müller, E. Yurchuk, P. Polakowski, J. Paul, D. Martin, T. Schenk, K. Khullar, A. Kersch, W. Weinreich, S. Riedel, K. Seidel, A. Kumar, T. M. Arruda, S. V. Kalinin, T. Schlosser, R. Boschke, R. van Bentum, U. Schroder, and T. Mikolajick. Ferroelectric hafnium oxide: A CMOS-compatible and highly scalable approach to future ferroelectric memories. pages 10.8.1–10.8.4. IEEE, December 2013.

[141] T. Schenk, S. Mueller, U. Schroeder, R. Materlik, A. Kersch, M. Popovici, C. Adelmann, S. Van Elshocht, and T. Mikolajick. Strontium doped hafnium oxide thin films: Wide process window for ferroelectric memories. In *Solid-State Device Research Conference (ESSDERC), 2013 Proceedings of the European*, pages 260–263. IEEE, 2013.

[142] J. Müller, T. S. Böscke, U. Schröder, Stefan Mueller, D. Bräuhaus, U. Böttger, Lothar Frey, and Thomas Mikolajick. Ferroelectricity in Simple Binary ZrO$_2$ and HfO$_2$. *Nano Letters*, 12(8):4318–4323, 2012.

[143] J. Müller, E. Yurchuk, T. Schlosser, J. Paul, R. Hoffmann, S. Müller, D. Martin, S. Slesazeck, P. Polakowski, J. Sundqvist, M. Czernohorsky, K. Seidel, P. Kücher, R. Boschke, M. Trentzsch, K. Gebauer, U. Schröder, and T. Mikolajick. Ferroelectricity in HfO$_2$ enables nonvolatile data storage in 28 nm HKMG. pages 25–26. IEEE, June 2012.

[144] P. Polakowski, S. Riedel, W. Weinreich, M. Rudolf, J. Sundqvist, K. Seidel, and J. Müller. Ferroelectric deep trench capacitors based on Al:HfO$_2$ for 3D nonvolatile memory applications. In *IEEE 6th International Memory Workshop (IMW)*, pages 1–4. IEEE, May 2014.

[145] S. Mueller, S. Slesazeck, T. Mikolajick, J. Müller P. Polakowski, and S. Flachowsky. Next-generation ferroelectric memories based on FE-HfO$_2$. In *Joint IEEE International Symposium on the Applications of Ferroelectric, International Symposium on Integrated Functionalities and Piezoelectric Force Microscopy Workshop (ISAF/ISIF/PFM)*, pages 233–236, Singapore, May 2015. IEEE.

[146] C.-H. Cheng and A. Chin. Low-Leakage-Current DRAM-Like Memory Using a One-Transistor Ferroelectric MOSFET With a Hf-Based Gate Dielectric. *IEEE Electron Device Letters*, 35(1):138–140, 2014.

[147] Y.-C. Chiu, C.-H. Cheng, C.-Y. Chang, M.-H. Lee, H.-H. Hsu, and S.-S. Yen. Low power 1T DRAM/NVM versatile memory featuring steep sub-60-mV/decade operation, fast

20-ns speed, and robust 85°C-extrapolated 10^{16} endurance. In *Symposium on VLSI Technology (VLSI Technology)*, pages T184–T185. IEEE, June 2015.

[148] M. D. Hanwell, D. E. Curtis, D. C. Lonie, T. Vandermeersch, E. Zurek, and G. R. Hutchison. Avogadro: an advanced semantic chemical editor, visualization, and analysis platform. *Journal of Cheminformatics*, 4(1):17, 2012.

[149] T. S. Böscke. *Crystalline Hafnia and Zirconia based Dielectrics for Memory Applications*. PhD thesis, Technische Universität Hamburg-Harburg, 2010.

[150] J. Müller. *Ferroelektrizität in Hafniumdioxid und deren Anwendung in nicht-flüchtigen Halbleiterspeichern*. PhD thesis, Technischen Universität Dresden, 2014.

[151] E. Yurchuk. *Electrical characterisation of ferroelectric field effect transistors based on ferroelectric HfO_2 thin films*. PhD thesis, Technische Universität Dresden, 2015.

[152] M. H. Park. *Novel material and device for ferroelectric memory: thin $Hf_{1-x}Zr_xO_2$ film and tri-states memory*. PhD thesis, Department of Materials Science and Engineering, College of Engineering, Seoul National University, 2014.

[153] S. F. Mueller. *Development of HfO_2-Based Ferroelectric Memories for Future CMOS Technology Nodes*. PhD thesis, Technische Universität Dresden, 2015.

[154] G. Friedbacher, H. Bubert, and Wiley InterScience (Online service). *Surface and thin film analysis: a compendium of principles, instrumentation, and applications*. Wiley-VCH Verlag, Weinheim, 2011.

[155] P. J. Potts, J.F. Bowles, S. .J. Reed, R. Cave, and Mineralogical Society (Great Britain), editors. *Microprobe techniques in the earth sciences*. Number 6 in Mineralogical Society series. Chapman & Hall, London; New York, 1st edition, 1995.

[156] D. K. Bowen and B. K Tanner. *X-ray metrology in semiconductor manufacturing*. CRC/Taylor & Francis, Boca Raton, 2006.

[157] J. Daillant and A. Gibaud, editors. *X-ray and neutron reflectivity: principles and applications*. Number 770 in Lecture notes in physics. Springer, Berlin, 2nd edition, 2009.

[158] Eric Lifshin, editor. *X-ray characterization of materials*. Wiley-VCH, Weinheim; New York, 1999.

[159] T. Schenk. Plasma Enhanced Atomic Layer Deposition and Characterization of Ferroelectric Aluminum Doped Hafnium Oxide, master thesis, Westsaxon UAS Zwickau. 2012.

[160] T. D. Huan, V. Sharma, G. A. Rossetti, and R. Ramprasad. Pathways towards ferroelectricity in hafnia. *Physical Review B*, 90(6), 2014.

[161] D. Damjanovic. Ferroelectric, dielectric and piezoelectric properties of ferroelectric thin films and ceramics. *Reports on Progress in Physics*, 61(9):1267–1324, 1998.

[162] S. Tiedke and T. Schmitz. Electrical Characterization of Nanoscale Ferroelectric Structures. In M. Alexe, A. Gruverman, P. Avouris, K. von Klitzing, H. Sakaki, and R. Wiesendanger, editors, *Nanoscale Characterisation of Ferroelectric Materials*, NanoScience and Technology. Springer Berlin Heidelberg, Berlin, Heidelberg, 2004.

[163] C. B. Sawyer and C. H. Tower. Rochelle salt as a dielectric. *Physical review*, 35(3):269, 1930.

[164] S. V. Kalinin and A. Gruverman, editors. *Scanning probe microscopy: electrical and electromechanical phenomena at the nanoscale*. Springer, New York, 2007.

[165] Piezoresponse force microscopy, Wikipedia – The Free Encyclopedia, https://en.wikipedia.org/wiki/Piezoresponse_force_microscopy (2015-11-27).

[166] N. Balke, S. Jesse, Q. Li, P. Maksymovych, M. B. Okatan, E. Strelcov, A. Tselev, and S. V. Kalinin. Current and surface charge modified hysteresis loops in ferroelectric thin films. *Journal of Applied Physics*, 118(7):072013, 2015

[167] S. Jesse and S. V. Kalinin. Band excitation in scanning probe microscopy: sines of change. *Journal of Physics D: Applied Physics*, 44(46):464006, 2011.

[168] S. Jesse, S. V. Kalinin, R. Proksch, A. P. Baddorf, and B. J. Rodriguez. The band excitation method in scanning probe microscopy for rapid mapping of energy dissipation on the nanoscale. *Nanotechnology*, 18(43):435503, 2007.

[169] S. Jesse, R. K. Vasudevan, L. Collins, E. Strelcov, M. B. Okatan, A. Belianinov, A. P. Baddorf, R. Proksch, and S. V. Kalinin. Band Excitation in Scanning Probe Microscopy: Recognition and Functional Imaging. *Annual Review of Physical Chemistry*, 65(1):519–536, 2014.

[170] N. Balke, P. Maksymovych, S. Jesse, I. I. Kravchenko Q. Li, and S. V. Kalinin. Exploring Local Electrostatic Effects with Scanning Probe Microscopy Implications for Piezoresponse Force Microscopy and Triboelectricity. *ACS Nano*, 8(10):10229–10236, 2014.

[171] X. Sang, E. D. Grimley, T. Schenk, U. Schroeder, and J. M. LeBeau. On the structural origins of ferroelectricity in HfO_2 thin films. *Applied Physics Letters*, 106(16):162905, 2015.

[172] M. Hoffmann, U. Schroeder, T. Schenk, T. Shimizu, H. Funakubo, O. Sakata, D. Pohl, M. Drescher, C. Adelmann, R. Materlik, A. Kersch, and T. Mikolajick. Stabilizing the ferroelectric phase in doped hafnium oxide. *Journal of Applied Physics*, 118(7):072006, 2015.

[173] Editorial. Beyond the diffraction limit. *Nature Photonics*, 3(7):361–361, 2009.

[174] D. B. Williams and C. B. Carter. *Transmission electron microscopy: a textbook for materials science.* Springer, New York, 2nd edition, 2009.

[175] A. Howie. Image Contrast And Localized Signal Selection Techniques. *Journal of Microscopy*, 117(1):11–23, 1979.

[176] D. A. Muller, N. Nakagawa, A. Ohtomo, J. L. Grazul, and H. Y. Hwang. Atomic-scale imaging of nanoengineered oxygen vacancy profiles in $SrTiO_3$. *Nature*, 430(7000):657–661, 2004.

[177] J. M. LeBeau, S. D. Findlay, X. Wang, A. J. Jacobson, L. J. Allen, and S. Stemmer. High-angle scattering of fast electrons from crystals containing heavy elements: Simulation and experiment. *Physical Review B*, 79(21), 2009.

[178] J. M. LeBeau, A. J. D'Alfonso, S. D. Findlay, S. Stemmer, and L. J. Allen. Quantitative comparisons of contrast in experimental and simulated bright-field scanning transmission electron microscopy images. *Physical Review B*, 80(17), 2009.

[179] S. D. Findlay, N. Shibata, H. Sawada, E. Okunishi, Y. Kondo, and Y. Ikuhara. Dynamics of annular bright field imaging in scanning transmission electron microscopy. *Ultramicroscopy*, 110(7):903–923, 2010.

[180] X. Sang and J. M. LeBeau. Revolving scanning transmission electron microscopy: Correcting sample drift distortion without prior knowledge. *Ultramicroscopy*, 138:28–35, 2014.

[181] F. Yang, F. Scheltens, D. McComb, D. B. Williams, and M. De Graef. Absorption Corrections for a Four-Quadrant SuperX EDS Detector. *Microscopy and Microanalysis*, 20(S3):100–101, 2014.

[182] J. M. LeBeau, S. D. Findlay, L. J. Allen, and S. Stemmer. Position averaged convergent beam electron diffraction: Theory and applications. *Ultramicroscopy*, 110(2):118–125, 2010.

[183] Evgenij Barsoukov and J. Ross Macdonald, editors. *Impedance spectroscopy: theory, experiment, and applications.* Wiley-Interscience, Hoboken, NJ, 2nd edition, 2005.

[184] Manual: Agilent 4294a, Agilent Technologies, http://literature.cdn.keysight.com/litweb/pdf/04294-90060.pdf?id=1000002189-1:epsg:man (2015-11-04). 2000.

[185] E. D. Grimley, T. Schenk, X. Sang, M. Pešić, U. Schroeder, T. Mikolajick, and J. M. LeBeau. Structural Changes Underlying Field-Cycling Phenomena in Ferroelectric HfO_2 Thin Films. *Advanced Electronic Materials*, 2(9):1600173, 2016.

[186] C. R. Pike, A. P. Roberts, and K. L. Verosub. Characterizing interactions in fine magnetic particle systems using first order reversal curves. *Journal of Applied Physics*, 85(9):6660, 1999.

[187] T. Schenk, M. Hoffmann, J. Ocker, M. Pešić, T. Mikolajick, and U. Schroeder. Complex Internal Bias Fields in Ferroelectric Hafnium Oxide. *ACS Applied Materials & Interfaces*, 7(36):20224–20233, 2015.

[188] M. Hoffmann. Electric Field Cycling Behavior of Ferroelectric HfO_2 Studied by First-Order Reversal Curves, project report, Faculty of Electrical and Computer Engineering, Institute of Semiconductors and Microsystems, Chair of Nanoelectronic Materials. 2015.

[189] aixACCT Systems. Manual: aixPlorer Software, Version 3.0.20.0 M1 – for TF Analyzer 3000 with FE-Module, 2013.

[190] M. Morozov. *Softening and hardening transitions in ferroelectric $Pb(Zr,Ti)O_3$ ceramics*. PhD thesis, École polytechnique fédérale de Lausanne, 2005.

[191] T. Shimizu, T. Yokouchi, T. Shiraishi, T. Oikawa, P. S. S. R. Krishnan, and H. Funakubo. Study on the effect of heat treatment conditions on metalorganic-chemical-vapor-deposited ferroelectric $Hf_{0.5}Zr_{0.5}O_2$ thin film on Ir electrode. *Japanese Journal of Applied Physics*, 53(9S):09PA04, September 2014.

[192] T. Shimizu, K. Katayama, T. Kiguchi, A. Akama, T. J. Konno, and H. Funakubo. Growth of epitaxial orthorhombic $YO_{1.5}$-substituted HfO_2 thin film. *Applied Physics Letters*, 107(3):032910, 2015.

[193] Y. Sharma, D. Barrionuevo, R. Agarwal, S. P. Pavunny, and R. S. Katiyar. Ferroelectricity in Rare-Earth Modified Hafnia Thin Films Deposited by Sequential Pulsed Laser Deposition. *ECS Solid State Letters*, 4(11):N13–N16, 2015.

[194] S. Starschich, D. Griesche, T. Schneller, R. Waser, and U. Böttger. Chemical solution deposition of ferroelectric yttrium-doped hafnium oxide films on platinum electrodes. *Applied Physics Letters*, 104(20):202903, 2014.

[195] S. Starschich, D. Griesche, T. Schneller, and U. Böttger. Chemical Solution Deposition of Ferroelectric Hafnium Oxide for Future Lead Free Ferroelectric Devices. *ECS Journal of Solid State Science and Technology*, 4(12):P419–P423, 2015.

[196] T. S. Böscke, St. Teichert, D. Bräuhaus, J. Müller, U. Schröder, U. Böttger, and T. Mikolajick. Phase transitions in ferroelectric silicon doped hafnium oxide. *Applied Physics Letters*, 99(11):112904, 2011.

[197] U. Schroeder, S. Mueller, J. Mueller, E. Yurchuk, D. Martin, C. Adelmann, T. Schloesser, R. van Bentum, and T. Mikolajick. Hafnium Oxide Based CMOS Compatible Ferroelectric Materials. *ECS Journal of Solid State Science and Technology*, 2(4):N69–N72, 2013.

[198] P. Polakowski and J. Müller. Ferroelectricity in undoped hafnium oxide. *Applied Physics Letters*, 106(23):232905, 2015.

[199] M. H. Park, H. J. Kim, Y. J. Ki, W. Lee, H. K. Kim, and C S Hwang. Effect of forming gas annealing on the ferroelectric properties of $Hf_{0.5}Zr_{0.5}O_2$ thin films with and without Pt electrodes. *Applied Physics Letters*, 102(11):112914, 2013.

[200] M. H. Park, H. J. Kim, Y. J. Kim, W. Jeon, T. Moon, and C. S. Hwang. Ferroelectric properties and switching endurance of $Hf_{0.5}Zr_{0.5}O_2$ films on TiN bottom and TiN or RuO_2 top electrodes:. *physica status solidi (RRL) - Rapid Research Letters*, 8(6):532–535, 2014.

[201] M. H. Park, H. J. Kim, Y. J. Kim, W. Lee, T. Moon, K. D. Kim, and C. S. Hwang. Study on the degradation mechanism of the ferroelectric properties of thin $Hf_{0.5}Zr_{0.5}O_2$ films on TiN and Ir electrodes. *Applied Physics Letters*, 105(7):072902, 2014.

[202] P. D. Lomenzo, P. Zhao, Q. Takmeel, S. Moghaddam, T. Nishida, M. Nelson, C. M. Fancher, E. D. Grimley, X. Sang, J. M. LeBeau, and J. L. Jones. Ferroelectric phenomena in Si-doped HfO2 thin films with TiN and Ir electrodes. *Journal of Vacuum Science & Technology B: Microelectronics and Nanometer Structures*, 32(3):03D123, 2014.

[203] P. D. Lomenzo, Q. Takmeel, C. Zhou, C. M. Fancher, E. Lambers, N. G. Rudawski, J. L. Jones, S. Moghaddam, and T. Nishida. TaN interface properties and electric field cycling effects on ferroelectric Si-doped HfO_2 thin films. *Journal of Applied Physics*, 117(13):134105, 2015.

[204] A. Mujica, A. Rubio, A. Munoz, and R. J. Needs. High-pressure phases of group-IV, III-V, and II-VI compounds. *Reviews of Modern Physics*, 75(3):863, 2003.

[205] R. J. Angel, J. Zhao, and N. L. Ross. General Rules for Predicting Phase Transitions in Perovskites due to Octahedral Tilting. *Physical Review Letters*, 95(2), 2005.

[206] P. Ayyub, V. R. Palkar, S. Chattopadhyay, and M. Multani. Effect of crystal size reduction on lattice symmetry and cooperative properties. *Physical Review B*, 51(9):6135, 1995.

[207] S. V. Ushakov, A. Navrotsky, Y. Yang, S. Stemmer, K. Kukli, M. Ritala, M. A. Leskelä, P. Fejes, A. Demkov, C. Wang, B.-Y. Nguyen, D. Triyoso, and P. Tobin. Crystallization in hafnia- and zirconia-based systems. *physica status solidi (b)*, 241(10):2268–2278, 2004.

[208] E. Roduner. Size matters: why nanomaterials are different. *Chemical Society Reviews*, 35(7):583, 2006.

[209] C. G Levi. Metastability and microstructure evolution in the synthesis of inorganics from precursors. *Acta Materialia*, 46(3):787–800, 1998.

[210] S. Stemmer, Z. Chen, C. G. Levi, P. S. Lysaght, B. Foran, J. A. Gisby, and J. R. Taylor. Application of metastable phase diagrams to silicate thin films for alternative gate dielectrics. *Japanese journal of applied physics*, 42(6R):3593, 2003.

[211] C. K. Ullal, K. R. Balasubramaniam, A. S. Gandhi, and V. Jayaram Non-equilibrium phase synthesis in Al_2O_3-Y_2O_3 by spray pyrolysis of nitrate precursors. *Acta materialia*, 49(14):2691–2699, 2001.

[212] Y. Kuroiwa, Y. Terado, S. J. Kim, A. Sawada, Y. Yamamura, S. Aoyagi, E. Nishibori, M. Sakata, and M. Takata. High-energy SR powder diffraction evidence of multisite disorder of Pb atom in cubic phase of $PbZr_{1-x}Ti_xO_3$. *Japanese Journal of Applied Physics*, 44(9S):7151, 2005.

[213] Y. Terado, S. J. Kim, C. Moriyoshi, Y. Kuroiwa, M. Iwata, and M. Takata. Disorder of Pb atom in cubic structure of $Pb(Zn_{1/3}Nb_{2/3})O_3$-$PbTiO_3$ system. *Japanese Journal of Applied Physics*, 45(9S):7552, 2006.

[214] C. Ederer and N. A. Spaldin. Influence of strain and oxygen vacancies on the magnetoelectric properties of multiferroic bismuth ferrite. *Physical Review B*, 71(22), 2005.

[215] R. Astala and P. D. Bristowe. Ab initio and classical simulations of defects in SrTiO3. *Computational Materials Science*, 22(1-2):81–86, 2001.

[216] H. Crogman and L. Bellaiche. Properties of vacancy-rich ordered (A,[ ])Nb_2O_6 perovskites. *Physical Review B*, 66(22), 2002.

[217] T. Chraska, A. H. King, and C. C. Berndt. On the size-dependent phase transformation in nanoparticulate zirconia. *Materials Science and Engineering: A*, 286(1):169–178, 2000.

[218] P. Li, I.-W. Chen, and J. E. Penner-Hahn. X-ray-absorption studies of zirconia polymorphs. II. Effect of Y_2O_3 dopant on ZrO_2 structure. *Physical Review B*, 48(14):10074, 1993.

[219] D. Sangalli and A. Debernardi. Exchange-correlation effects in the monoclinic to tetragonal phase stabilization of yttrium-doped ZrO_2: A first-principles approach. *Physical Review B*, 84(21), 2011.

[220] P. Li, I-W. Chen, and J. E. Penner-Hahn. Effect of Dopants on Zirconia Stabilization-An X-ray Absorption Study: I, Trivalent Dopants. *Journal of the American Ceramic Society*, 77(1):118–128, 1994.

[221] A. Navrotsky. Thermochemical insights into refractory ceramic materials based on oxides with large tetravalent cations. *Journal of Materials Chemistry*, 15(19):1883, 2005.

[222] S. Aramaki and R. Roy. Revised Phase Diagram for the System Al_2O_3-SiO_2. *Journal of the American Ceramic Society*, 45(5):229–242, 1962.

[223] H. P. Pinto, R. M. Nieminen, and S. D. Elliott. Ab initio study of γ-Al_2O_3 surfaces. *Physical Review B*, 70(12), 2004.

[224] R. M. Hazen and L. W. Finger. Crystal structure and compressibility of zircon at high pressure. *American Mineralogist*, 64(1-2):196–201, 1979.

[225] H. Fujimori, M. Yashima, S. Sasaki, M. Kakihana, T. Mori, M. Tanaka, and M. Yoshimura. Cubic-tetragonal phase change of yttria-doped hafnia solid solution: high-resolution X-ray diffraction and Raman scattering. *Chemical Physics Letters*, 346(3-4):217–223, 2001.

[226] M. Yashima, S. Sasaki, Y. Yamaguchi, M. Kakihana, M. Yoshimura, and T. Mori. Internal distortion in ZrO_2-CeO_2 solid solutions: Neutron and high-resolution synchrotron x-ray diffraction study. *Applied Physics Letters*, 72(2):182, 1998.

[227] R. D. Shannon. Revised effective ionic radii and systematic studies of interatomic distances in halides and chalcogenides. *Acta Crystallographica Section A*, 32(5):751–767, 1976.

[228] Database of Ionic Radii, Atomistic Simulation Group, Department of Materials, Imperial College London, http://abulafia.mt.ic.ac.uk/shannon/ (2015-10-12). 2015.

[229] Gregory S. Rohrer. *Structure and bonding in crystalline materials*. Cambridge University Press, Cambridge ; New York, 2001.

[230] D. Cunningham. A First-Principles Examination of Dopants in HfO_2, Honors Scholar Theses, Paper 359, University of Connecticut, Department of Materials Science and Engineering, http://digitalcommons.uconn.edu/srhonors_theses/359. 2014.

[231] L. Zhao, M. Nelson, H. Aldridge, T. Iamsasri, C. M. Fancher, J. S. Forrester, T. Nishida, S. Moghaddam, and J. L. Jones. Crystal structure of Si-doped HfO_2. *Journal of Applied Physics*, 115(3):034104, 2014.

[232] L. Zhao, D. Hou, T.-M. Usher, T. Iamsasri, C. M. Fancher, J. S. Forrester, T. Nishida, S. Moghaddam, and J. L. Jones. Structure of 3 at.% and 9 at.% Si-doped HfO_2 from combined refinement of X-ray and neutron diffraction patterns. *Journal of Alloys and Compounds*, 646:655–661, 2015.

[233] C. M. Fancher, L. Zhao, M. Nelson, L. Bai, G. Shen, and J. L. Jones. Pressure-induced structures of Si-doped HfO_2. *Journal of Applied Physics*, 117(23):234102, 2015.

[234] D. Hou, C. M. Fancher, L. Zhao, G. Esteves, and J. L. Jones. Processing and crystallographic structure of non-equilibrium Si-doped HfO_2. *Journal of Applied Physics*, 117(24):244103, 2015.

[235] S. Clima, D. J. Wouters, C. Adelmann, T. Schenk, U. Schroeder, M. Jurczak, and G. Pourtois. Identification of the ferroelectric switching process and dopant-dependent switching properties in orthorhombic HfO_2: A first principles insight. *Applied Physics Letters*, 104(9):092906, 2014.

[236] S. E. Reyes-Lillo, K. F. Garrity, and K. M. Rabe. Antiferroelectricity in thin-film ZrO_2 from first principles. *Physical Review B*, 90(14), 2014.

[237] S. Starschich, C. Künneth, R. Materlik, U. Schroeder, A. Kersch, and U. Böttger. Ferroelectricity in $Hf_{1-x}Zr_xO_2$ and pure ZrO_2 induced by the grain size effect. *Applied Physics Letters (in review)*, 2016.

[238] J. Müller, T. S. Böscke, D. Bräuhaus, U. Schröder, U. Böttger, J. Sundqvist, P. Kücher, T. Mikolajick, and L. Frey. Ferroelectric $Zr_{0.5}Hf_{0.5}O_2$ thin films for nonvolatile memory applications. *Applied Physics Letters*, 99(11):112901, 2011.

[239] S. Mueller, J. Müller, R. Hoffmann, E. Yurchuk, T. Schlosser, R. Boschke, J. Paul, M. Goldbach, T. Herrmann, A. Zaka, U. Schröder, and T. Mikolajick. From MFM Capacitors Toward Ferroelectric Transistors: Endurance and Disturb Characteristics of HfO_2-Based FeFET Devices. *IEEE Transactions on Electron Devices*, 60(12):4199–4205, 2013.

[240] S. Mueller, J. Müller, U. Schröder, and T. Mikolajick Reliability Characteristics of Ferroelectric Si:HfO_2 Thin Films for Memory Applications. *IEEE Transactions on Device and Materials Reliability*, 13(1):93–97, 2013.

[241] S. Mueller, E. Yurchuk, S. Slesazeck, T. Mikolajick, J. Müller, T. Herrmann, and A. Zaka. Performance investigation and optimization of Si:HfO_2 FeFETs on a 28 nm bulk technology. pages 248–251. IEEE, 2013.

[242] J. Müller, T. S. Böscke, U. Schröder, R. Hoffmann, T. Mikolajick, and L. Frey. Nanosecond Polarization Switching and Long Retention in a Novel MFIS-FET Based on Ferroelectric HfO_2. *IEEE Electron Device Letters*, 33(2):185–187, 2012.

[243] D. Martin, J. Müller, T. Schenk, T. M. Arruda, A. Kumar, E. Strelcov, E. Yurchuk, S. Müller, D. Pohl, U. Schröder, S. V. Kalinin, and T. Mikolajick. Ferroelectricity in Si-Doped HfO_2 Revealed: A Binary Lead-Free Ferroelectric. *Advanced Materials*, 26(48):8198–8202, 2014.

[244] T. S. Böscke, J. Müller, D. Bräuhaus, U. Schröder, and U. Böttger. Ferroelectricity in hafnium oxide: CMOS compatible ferroelectric field effect transistors. pages 24.5.1–24.5.4. IEEE, December 2011.

[245] M. H. Park, H. J. Kim, Y. J. Kim, W. Lee, T. Moon, and C. S. Hwang. Evolution of phases and ferroelectric properties of thin $Hf_{0.5}Zr_{0.5}O_2$ films according to the thickness and annealing temperature. *Applied Physics Letters*, 102(24):242905, 2013.

[246] P. D. Lomenzo, Q. Takmeel, C. Zhou, C.-C. Chung, S. Moghaddam, J. L. Jones, and T. Nishida. Mixed Al and Si doping in ferroelectric HfO_2 thin films. *Applied Physics Letters*, 107(24):242903, 2015.

[247] A. N. Norris, P. Sheng, and A. J. Callegari. Effective-medium theories for two-phase dielectric media. *Journal of Applied Physics*, 57(6):1990, 1985.

[248] A. H. Sihvola and J. A. Kong. Effective permittivity of dielectric mixtures. *IEEE Transactions on Geoscience and Remote Sensing*, 26(4):420–429, 1988.

[249] T. C. Choy. *Effective medium theory;principles and applications*. Oxford University Press, Oxford, 2015.

[250] E. Yurchuk, J. Müller, S. Knebel, J. Sundqvist, A. P. Graham, T. Melde, U. Schröder, and T. Mikolajick. Impact of layer thickness on the ferroelectric behaviour of silicon doped hafnium oxide thin films. *Thin Solid Films*, 533:88–92, 2013.

[251] P. D. Lomenzo, Q. Takmeel, S. Moghaddam, and T. Nishida. Annealing behavior of ferroelectric Si-doped HfO_2 thin films. *Thin Solid Films*, 615:139–144, 2016.

[252] M. M. Frank. High-k/metal gate innovations enabling continued CMOS scaling. In *Proceedings of the IEEE ESSCIRC*, pages 50–58. IEEE, 2011.

[253] J. M. Lopes, E. Durğun Özben, M. Schnee, R. Luptak, A. Nichau, A. Tiedemann, W. Yu, Q.-T. Zhao, A. Besmehn, U. Breuer, M. Luysberg, S. Lenk, J. Schubert, and . Mantl. Electrical and Structural Properties of Ternary Rare-Earth Oxides on Si and Higher Mobility Substrates and their Integration as High-k Gate Dielectrics in MOSFET Devices. *ECS Transactions*, 35(4):461–479, 2011.

[254] M. Hoffmann, T. Schenk, I. Kulemanov, C. Adelmann, M. Popovici, U. Schroeder, and T. Mikolajick. Low Temperature Compatible Hafnium Oxide Based Ferroelectrics. *Ferroelectrics*, 480(1):16–23, 2015.

[255] H.-J. Bunge and P. R. Morris. *Texture analysis in materials science: mathematical methods*. Elsevier Science, 1982.

[256] L. Lutterotti, H.-R. Wenk, S. Matthies, M. Ferrari, N. C. Popa, R. W. Grosse-Kunstleve, L. Cont, and M. Bortolotti. MAUD - Material Analysis Using Diffraction, http://maud.radiographema.eu/.

[257] L. Lutterotti, M. Bortolotti, G. Ischia, I. Lonardelli, and H. R. Wenk. Rietveld texture analysis from diffraction images. *Z. Kristallogr. Suppl*, 26:125–130, 2007.

[258] I. C. Noyan, T. C. Huang, and B. R. York. Residual stress/strain analysis in thin films by X-ray diffraction. *Critical Reviews in Solid State and Materials Sciences*, 20(2):125–177, 1995.

[259] K. A. Müller and H. Burkard. SrTiO$_3$: An intrinsic quantum paraelectric below 4 K. *Physical Review B*, 19(7):3593–3602, 1979.

[260] W. Zhong and D. Vanderbilt. Effect of quantum fluctuations on structural phase transitions in SrTiO$_3$ and BaTiO$_3$. *Physical Review B*, 53(9):5047–5050, 1996.

[261] T. Schneider, H. Beck, and E. Stoll. Quantum effects in an n -component vector model for structural phase transitions. *Physical Review B*, 13(3):1123–1130, 1976.

[262] H. Bilz, A. Bussmann, G. Benedek, H. Büttner, and D. Strauch. Microscopic model of ferroelectric soft modes. *Ferroelectrics*, 25(1):339–342, 1980.

[263] J. H. Barrett. Dielectric Constant in Perovskite Type Crystals. *Physical Review*, 86(1):118–120, 1952.

[264] Y. S. Kim, J. Kim, S. J. Moon, W. S. Choi, Y. J. Chang, J.-G. Yoon, J. Yu, J.-S. Chung, and T. W. Noh. Localized electronic states induced by defects and possible origin of ferroelectricity in strontium titanate thin films. *Applied Physics Letters*, 94(20):202906, 2009.

[265] Y. S. Kim, D. J. Kim, T. H. Kim, T. W. Noh, J. S. Cho, B. H. Park, and J.-G. Yoon. Observation of room-temperature ferroelectricity in tetragonal strontium titanate thin films on SrTiO$_3$ (001) substrates. *Applied Physics Letters*, 91(4):042908, 2007.

[266] M. Takesada, M. Itoh, and T. Yagi. Perfect Softening of the Ferroelectric Mode in the Isotope-Exchanged Strontium Titanate of SrTiO$_3^{18}$ Studied by Light Scattering. *Physical Review Letters*, 96(22), 2006.

[267] M. Itoh, R. Wang, Y. Inaguma, T. Yamaguchi, Y-J. Shan, and T. Nakamura. Ferroelectricity Induced by Oxygen Isotope Exchange in Strontium Titanate Perovskite. *Physical Review Letters*, 82(17):3540–3543, 1999.

[268] D. Damjanovic. Hysteresis in Piezoelectric and Ferroelectric Materials. In G. Bertotti and I. D. Mayergoyz, editors, *The Science of Hysteresis: Hysteresis in materials*, volume 3. Academic, Amsterdam; Boston, 1st edition, 2006

[269] S. Knebel, S. Kupke, U. Schroeder, S. Slesazeck, T. Mikolajick, R. Agaiby, and M. Trentzsch. Influence of Frequency Dependent Time to Breakdown on High-K/Metal Gate Reliability. *IEEE Transactions on Electron Devices*, 60(7):2368–2371, 2013.

[270] G. Bersuker, D. Heh, C. Young, H. Park, P. Khanal, L. Larcher, A. Padovani, P. Lenahan, J. Ryan, B. H. Lee, H. Tseng, and R. Jammy. Breakdown in the metal/high-k gate stack: Identifying the "weak link" in the multilayer dielectric. *IEDM Tech. Dig.*, pages 791–794, 2008.

[271] W. L. Warren, Karel Vanheusden, D. Dimos, G. E. Pike, and B. A. Tuttle. Oxygen Vacancy Motion in Perovskite Oxides. *Journal of the American Ceramic Society*, 79(2):536–538, 1996.

[272] U. Robels, L. Schneider-Stormann, and G. Arlt. Dielectric aging and its temperature dependence in ferroelectric ceramics. *Ferroelectrics*, 168(1):301–311, 1995.

[273] E. Erdem, R.-A. Eichel, Cs. Fetzer, I. Dézsi, S. Lauterbach, H.-J. Kleebe, and A. G. Balogh. Site of incorporation and solubility for Fe ions in acceptor-doped PZT ceramics. *Journal of Applied Physics*, 107(5):054109, 2010.

[274] R. D. Klissurska, K. G. Brooks, and N. Setter. Effect of dopants on the crystallization mechanism of PZT thin films. *Ferroelectrics*, 225(1):327–334, 1999.

[275] J. L. Gavartin, D. Muñoz Ramo, A. L. Shluger, G. Bersuker, and B. H. Lee. Negative oxygen vacancies in HfO_2 as charge traps in high-k stacks. *Applied Physics Letters*, 89(8):082908, 2006.

[276] Y. Park, K. W. Jeong, and J. T. Song. Effect of excess Pb on fatigue properties of PZT thin films prepared by rf-magnetron sputtering. *Materials Letters*, 56(4):481–485, 2002.

[277] I. Shturman, G. E. Shter, A. Etin, and G. S. Grader. Effect of $LaNiO_3$ electrodes and lead oxide excess on chemical solution deposition derived $Pb(Zr_x,Ti_{1-x})O_3$ films. *Thin Solid Films*, 517(8):2767–2774, 2009.

[278] W. Weinreich, R. Reiche, M. Lemberger, G. Jegert, J. Müller, L. Wilde, S. Teichert, J. Heitmann, E. Erben, L. Oberbeck, U. Schröder, A. J. Bauer, and H. Ryssel. Impact of interface variations on J-V and C-V polarity asymmetry of MIM capacitors with amorphous and crystalline $Zr_{(1-x)}Al_xO_2$ films. *Microelectronic Engineering*, 86(7-9):1826–1829, 2009.

[279] M. H. Park, H. J. Kim, Y. J. Kim, T. Moon, K. D. Kim, and C. S. Hwang. Thin $Hf_xZr_{1-x}O_2$ Films: A New Lead-Free System for Electrostatic Supercapacitors with Large Energy Storage Density and Robust Thermal Stability. *Advanced Energy Materials*, 4(16):1400610, 2014.

[280] S. Starschich, S. Menzel, and U. Böttger. Evidence for oxygen vacancies movement during wake-up in ferroelectric hafnium oxide. *Applied Physics Letters*, 108(3):032903, 2016.

[281] C. Li, Y. Yao, X. Shen, Y. Wang, J. Li, C. Gu, R. Yu, Q. Liu, and M. Liu. Dynamic observation of oxygen vacancies in hafnia layer by in situ transmission electron microscopy. *Nano Research*, 8(11):3571–3579, 2015.

[282] M. Pešić, F. P. G. Fengler, L. Larcher, A. Padovani, T. Schenk, E. D. Grimley, X. Sang, J. M. LeBeau, S. Slesazeck, U. Schroeder, and T. Mikolajick. Physical Mechanisms behind the Field-Cycling Behavior of HfO_2-Based Ferroelectric Capacitors. *Advanced Functional Materials*, 26(25):4601–4612, 2016.

[283] N. Capron, P. Broqvist, and A. Pasquarello. Migration of oxygen vacancy in HfO_2 and across the HfO_2/SiO_2 interface: A first-principles investigation. *Applied Physics Letters*, 91(19):192905, 2007.

[284] S. Kupke. *Reliability of high-k / metal gate field-effect transistors considering circuit operational constraints*. Research at namlab. BoD - Books on Demand, Norderstedt, 2016.

[285] L. Pintilie. Charge Transport in Ferroelectric Thin Films. In *Ferroelectrics - physical effects*. InTech, Rijeka, 2011.

[286] N. Balke, P. Maksymovych, S. Jesse, A. Herklotz, A. Tselev, C.-B. Eom, I. I. Kravchenko, P. Yu, and S. V. Kalinin. Differentiating Ferroelectric and Nonferroelectric Electromechanical Effects with Scanning Probe Microscopy. *ACS Nano*, 9(6):6484–6492, 2015.

[287] E. Strelcov, Y. Kim, J. C. Yang, Y. H. Chu, P. Yu, X. Lu, S. Jesse, and S. V. Kalinin. Role of measurement voltage on hysteresis loop shape in Piezoresponse Force Microscopy. *Applied Physics Letters*, 101(19):192902, 2012.

[288] A. Chernikova, M. Kozodaev, A. Markeev, D. Negrov, M. Spiridonov, S. Zarubin, O. Bak, P. Buragohain, H. Lu, E. Suvorova, A. Gruverman, and A. Zenkevich. Ultrathin $Hf_{0.5}Zr_{0.5}O_2$ Ferroelectric Films on Si. *ACS Applied Materials & Interfaces*, 8(11):7232–7237, 2016.

[289] C.-H. Kim, S.-I. Pyun, and J.-H. Kim. An investigation of the capacitance dispersion on the fractal carbon electrode with edge and basal orientations. *Electrochimica Acta*, 48(23):3455–3463, 2003.

[290] M. E. Orazem, I. Frateur, B. Tribollet, V. Vivier, S. Marcelin, N. Pébere, A. L. Bunge, E. A. White, D. P. Riemer, and M. Musiani. Dielectric Properties of Materials Showing Constant-Phase-Element (CPE) Impedance Response. *Journal of the Electrochemical Society*, 160(6):C215–C225, 2013.

[291] M. Musiani, M. E. Orazem, N. Pébère, B. Tribollet, and V. Vivier. Constant-Phase-Element Behavior Caused by Coupled Resistivity and Permittivity Distributions in Films. *Journal of The Electrochemical Society*, 158(12):C424, 2011.

[292] B. Hirschorn, M. E. Orazem, B. Tribollet, V. Vivier, I. Frateur, and M. Musiani. Constant-Phase-Element Behavior Caused by Resistivity Distributions in Films. *Journal of The Electrochemical Society*, 157(12):C452, 2010.

[293] B. Hirschorn, M. E. Orazem, B. Tribollet, V. Vivier, I. Frateur, and M. Musiani. Constant-Phase-Element Behavior Caused by Resistivity Distributions in Films. *Journal of The Electrochemical Society*, 157(12):C458, 2010.

[294] J.-B. Jorcin, M. E. Orazem, N. Pébère, and B. Tribollet. CPE analysis by local electrochemical impedance spectroscopy. *Electrochimica Acta*, 51(8-9):1473–1479, 2006.

[295] V. V. Afanas'ev, A. Stesmans, L. Pantisano, S. Cimino, C. Adelmann, L. Goux, Y. Y. Chen, J. A. Kittl, D. Wouters, and M. Jurczak. TiN_x/HfO_2 interface dipole induced by oxygen scavenging. *Applied Physics Letters*, 98(13):132901, 2011.

[296] W. Weinreich, A. Shariq, K. Seidel, J. Sundqvist, A. Paskaleva, M. Lemberger, and A. J. Bauer. Detailed leakage current analysis of metal-insulator-metal capacitors with ZrO_2, $ZrO_2/SiO_2/ZrO_2$, and $ZrO_2/Al_2O_3/ZrO_2$ as dielectric and TiN electrodes. *Journal of Vacuum Science & Technology B: Microelectronics and Nanometer Structures*, 31(1):01A109, 2013.

A Appendix: Dopant Overview from J. Müller's Dissertation

The table below and corresponding references are translated from Johannes Müller's dissertation[150] (Tabelle 3.1) by the author of the present work.

Literature overview of experimental and theoretical high-temperature stability in doped HfO_2 systems. r_{ion} denotes the ionic radius by Shannon [142] for a typical coordination number. The * marks simulation results which are only of theoretical relevance because a phase segregation cannot be excluded in practice.

Ion	r_{ion} in Å (coordination)	Phase Stabilization in HfO_2 Experiment	Ab-initio Simulation
Hf^{4+}	0.83 (VIII)	mainly monoclinic [143, 144]	monoclinic [145, 146]
Al^{3+}	0.39 (IV)	tetragonal > 7 – 14 mol% $AlO_{1.5}$ [147, 148]	tetragonal > 11.1 mol% $AlO_{1.5}$ [149]
C^{4+}	0.29 (IV)	tetragonal (grain size reduction) C often impurity [150]	inactive [145, 146]
Ca^{2+}	1.12 (VIII)	cubic > 6 mol% CaO [115, 151]	no data
Ce^{4+}	0.97 (VIII)	cubic or tetragonal > 10 mol% CeO_2 [152, 153]	* tetragonal > 21 mol% CeO_2 [145, 146]
Dy^{3+}	1.03 (VIII)	tetragonal → cubic > 10 mol% $DyO_{1.5}$ [154–156]	no data
Er^{3+}	1.00 (VIII)	tetragonal → cubic > 10 mol% $ErO_{1.5}$ [156–159]	no data
Gd^{3+}	1.05 (VIII)	tetragonal → cubic > 10 – 20 mol% $GdO_{1.5}$ [156, 160–164]	cubic > 9.9 mol% $GdO_{1.5}$ [149]
Ge^{4+}	0.39 (IV)	tetragonal > 15 – 18 mol% GeO_2 [125, 165]	tetragonal > 7.5 – 18.6 mol% GeO_2 [145, 146, 149]
La^{3+}	1.16 (VIII)	cubic > 4 mol% $LaO_{1.5}$ [133, 166–169]	no data
Sc^{3+}	0.87 (VIII)	cubic > 12 mol% $ScO_{1.5}$ [154, 155]	cubic > 12.8 mol% $ScO_{1.5}$ [149]
Si^{4+}	0.26 (IV)	tetragonal > 5 – 9 mol% SiO_2 [132, 163, 166, 169–173]	tetragonal > 6.5 – 10.9 mol% SiO_2 [145, 149, 174]
Sn^{4+}	0.55 (IV)	no data	* tetragonal > 19 mol% SnO_2 [145, 146, 149]
Ti^{4+}	0.42 (IV)	inactive [175]	marginal [145, 146]
Y^{3+}	1.02 (VIII)	tetragonal → cubic > 2.5 – 13 mol% $YO_{1.5}$ [29, 128–141]	cubic > 10.6 mol% $YO_{1.5}$ [149]
Zr^{4+}	0.84 (VIII)	monoclinic or tetragonal, gradual transition [120, 124, 143, 176–178]	inactive [149] (surface energy not considered)

References (as provided)

[115] Curtis, C. E., L. M. Doney and J. R. Johnson: *Some properties of hafnium oxide, hafnium silicate, calcium hafnate, and hafnium carbide.* Journal of the American Ceramic Society, 37(10):458–465, 1954.

[125] Lefevre, J.: Flourite-type structural phase modifications in systems having a zirconioum or hafnium oxide base. Annales de chimie, 8:117–149, 1963.

[132] Ushakov, S. V., C. E. Brown, A. Navrotsky, A. Demkov, C. Wang and B.-Y. Nguyen: *Thermal analyses of bulk amorphous oxides and silicates of zirconium and hafnium.* Mater. Res. Soc. Symp. Proc. (Materials Research Society Proceedings), 745:N1.4.1–N1.4.6, 2003.

[133] Ushakov, S. V., C. E. Brown and A. Navrotsky: *Effect of La and Y on crystallization temperatures of hafnia and zirconia.* Journal of Materials Research, 19(3):693–696, 2004.

[143] Triyoso, D. H., R. I. Hegde, J. K. Schaeffer, D. Roan, P. J. Tobin, S. B. Samavedam, B. E. Jr. White, R. Gregory and X. D.Wang: *Impact of Zr addition on properties of atomic layer deposited HfO_2.* Applied Physics Letters, 88(22):222901, 2006.

[144] Hackley, J. C. and T. Gougousi: *Properties of atomic layer deposited HfO_2 thin films.* Thin Solid Films, 517(24):6576–6583, 2009.

[145] Fischer, D. and A. Kersch: *Stabilization of the high-k tetragonal phase in HfO_2: the influence of dopants and temperature from ab initio simulations.* Journal of Applied Physics, 104(8):084104, 2008.

[146] Fischer, D. and A. Kersch: *The effect of dopants on the dielectric constant of HfO_2 and ZrO_2 from first principles.* Applied Physics Letters, 92(1):012908, 2008.

[147] Yang, Y., W. Zhu, T. P. Ma and S. Stemmer: *High-temperature phase stability of hafnium aluminate films for alternative gate dielectrics.* Journal of Applied Physics, 95(7), 2004.

[148] Park, P. K. and S.-W. Kang: *Enhancement of dielectric constant in HfO_2 thin films by the addition of Al_2O_3.* Applied Physics Letters, 89(19):192905, 2006.

[149] Lee, C.-K., E. Cho, H.-S. Lee, C. S. Hwang and S. Han: *First-principles study on doping and phase stability of HfO_2.* Physical Review B, 78(1):012102, 2008.

[150] Cho, D.-Y., H. S. Jung, I.-H. Yu, J. H. Yoon, H. K. Kim, S. Y. Lee, S. H. Jeon, S. Han, J. H. Kim, T. J. Park, B.-G. Park and C. S. Hwang: *Stabilization of tetragonal HfO_2 under low active oxygen source environment in atomic layer deposition.* Chemistry of Materials, 24(18):3534–3543, 2012.

[151] Barolin, S. A., M. C. Caracoche, J. A. Martínez, P. C. Rivas, M. A. Taylor, A. F. Pasquevich and O. A. de Sanctis: *Thermal evolution of CaO-doped HfO_2 films and powders.*Journal of Physics: Conference Series, 167(1):012052, 2009.

[152] Chalker, P. R., M. Werner, S. Romani, R. J. Potter, K. Black, H. C. Aspinall, A. C. Jones, C. Z. Zhao, S. Taylor and P. N. Heys: *Permittivity enhancement of hafnium dioxide high-kappa films by cerium doping.* Applied Physics Letters, 93(18):182911, 2008.

[153] Kim, W.-H., M.-K. Kim, I.-K. Oh, W. J. Maeng, T. Cheon, S.-H. Kim, A. Noori, D. Thompson, S. Chu and H. Kim: *Significant enhancement of the dielectric constant through the doping of CeO_2 into HfO_2 by atomic layer deposition.* Journal of the American Ceramic Society, 97(4):1164–1169, 2014.

[154] Adelmann, C., V. Sriramkumar, S. van Elshocht, P. Lehnen and T. Conard: *Dielectric properties of dysprosium- and scandium-doped hafnium dioxide thin films.* Applied Physics Letters, 91(16):162902, 2007.

[155] Adelmann, C., P. Lehnen, S. van Elshocht, C. Zhao, B. Brijs, A. Franquet, T. Conard, M. Roeckerath, J. Schubert, O. Boissière and C. Lohe: *Growth of dysprosium-, scandium-, and hafnium-based third generation high-k dielectrics by atomic vapor deposition.* Chemical Vapor Deposition, 13(10):567–573, 2007.

[156] Govindarajan, S., T. S. Böscke, P. Sivasubramani, P. D. Kirsch, B. H. Lee, H. H Tseng, R. Jammy, U. Schröder, S. Ramanathan and B. E. Gnade: *Higher permittivity rare earth doped HfO_2 for sub-45-nm metal-insulator-semiconductor devices.* Applied Physics Letters, 91(6):062906, 2007.

[157] Wiemer, C., L. Lamagna, S. Baldovino, M. Perego, S. Schamm-Chardon, P. E. Coulon, O. Salicio, G. Congedo, S. Spiga and M. Fanciulli: *Dielectric properties of Er-doped HfO2 (Er 15%) grown by atomic layer deposition for high-kappa gate stacks.* Applied Physics Letters, 96(18):182901, 2010.

[158] Duran, P., C. Pascua, J.-P. Coutures and S. R. Skaggs: *Phase relations and ordering in the system erbia-hafnia.* Journal of the American Ceramic Society, 66(2):101–106, 1983.

[159] Toomey, B., K. Cherkaoui, S. Monaghan, V. Djara, E. O'Connor, D. O'Connell, L. Oberbeck, E. Tois, T. Blomberg, S. B. Newcomb and P. K. Hurley: *The structural and electrical characterization of a HfErOx dielectric for MIM capacitor DRAM applications.* Microelectronic Engineering, 94:7–10, 2012.

[160] Losovyj, Y. B., I. Ketsman, A. Sokolov, K. D. Belashchenko, P. A. Dowben, J. Tang and Z. Wang: *The electronic structure change with Gd doping of HfO_2 on silicon.* Applied Physics Letters, 91(13):132908, 2007.

[161] Yashima, M., H. Takahashi, K. Ohtake, T. Hirose, M. Kakihana, H. Arashi, Y. Ikuma, Y. Suzuki and M. Yoshimura: *Formation of metastable forms by quenching of the HfO_2-$RO_{1.5}$.* J. Phys. Chem. Solids (Journal of Physics and Chemistry of Solids), 57(3):289–295, 1996.

[162] Govindarajan, S., T. S. Boescke, P. D. Kirsch, M. A. Quevedo-Lopez, P. Sivasubramani, S. C. Song, R. W. Wallace, B. E. Gnade, P. Y. Hung, J. Price, U. Schroeder, S. Ramanathan, B. H. Lee and R. Jammy: *Higher permittivity rare earth-doped HfO_2 and ZrO_2 dielectrics for logic and memory applications.* In: *International Symposium on VLSI Technology, Systems and Applications (VLSI-TSA)*, pp. 1–2, 2007.

[163] Böscke, T. S., S. Govindarajan, C. Fachmann, J. Heitmann, A. Avelan, U. Schröder, S. Kudelka, P. D. Kirsch, C. Krug, P. Y. Hung, S. C. Song, B. S. Ju, J. Price, G. Pant, B. E. Gnade, W. Krautschneider, B.-H Lee and R. Jammy: *Tetragonal phase stabilization by doping as an enabler of thermally stable HfO_2 based MIM and MIS capacitors for sub 50 nm deep trench DRAM.* In: *IEEE International Electron Devices Meeting (IEDM)*, pp. 1–4, 2006.

[164] Dole, S. L., O. Hunter and F. W. Calderwood: *Elastic properties of stabilized HfO_2 compositions.* Journal of the American Ceramic Society, 63(3-4):136–139, 1980.

[165] Miotti, L., K. P. Bastos, G. Lucovsky, C. Radtke and D. Nordlund: *Ge doped HfO_2 thin films investigated by x-ray absorption spectroscopy.* Journal of Vacuum Science & Technology A: Vacuum, Surfaces, and Films, 28(4):693–696, 2010.

[166] Ushakov, S. V., A. Navrotsky, Y. Yang, S. Stemmer, K. Kukli, M. Ritala, M. A. Leskelä, P. Fejes, A. Demkov, C. Wang, B. Y Nguyen, D. Triyoso and P. Tobin: *Crystallization in hafnia- and zirconia-based systems.* physica status solidi (b), 241(10):2268–2278, 2004.

[167] He, W., D. S. H. Chan, S.-J. Kim, Y.-S. Kim, S.-T Kim and B. J. Cho: *Process and material properties of $HfLaO_x$ prepared by atomic layer deposition.* Journal of The Electrochemical Society, 155(10):G189–G193, 2008.

[168] Yamamoto, Y., K. Kita, K. Kyuno and A. Toriumi: *Structural and electrical properties of $HfLaO_x$ films for an amorphous high-k gate insulator.* Applied Physics Letters, 89(3):032903, 2006.

[169] Cao, D., X. Cheng, Y. Yu, X. Li, C. Liu, D. Shen and S. Mandl: *Competitive Si and La effect in HfO_2 phase stabilization in multi-layer $(La_2O_3)0.08(HfO_2)$ films.* Applied Physics Letters, 103(8):081607, 2013.

[170] Tomida, K., K. Kita and A. Toriumi: *Dielectric constant enhancement due to Si incorporation into HfO_2.* Applied Physics Letters, 89(14):142902, 2006.

[171] Böscke, T. S., S. Govindarajan, P. D. Kirsch, P. Y. Hung, C. Krug, B. H. Lee, J. Heitmann, U. Schröder, G. Pant, B. E. Gnade and W. H. Krautschneider: *Stabilization of higher-kappa tetragonal HfO_2 by SiO_2 admixture enabling thermally stable metal-insulator-metal capacitors*. Applied Physics Letters, 91(7):072902, 2007.

[172] Fachmann, C., L. Frey, S. Kudelka, T. Boescke, S. Nawka, E. Erben and T. Doll: Tuning the dielectric properties of hafnium silicate films. Microelectronic Engineering, 84(12):2883–2887, 2007.

[173] Neumayer, D. A. and E. Cartier: *Materials characterization of ZrO_2-SiO_2 and HfO_2-SiO_2 binary oxides deposited by chemical solution deposition by chemical solution deposition*. Journal of Applied Physics, 90(4):1801–1808, 2001.

[174] Fischer, D. and A. Kersch: *Ab initio study of high permittivity phase stabilization in HfSiO*. Microelectronic Engineering, 84(9-10):2039–2042, 2007.

[175] Triyoso, D. H., R. I. Hegde, S. Zollner, M. E. Ramon, S. Kalpat, R. Gregory, X.-D Wang, J. Jiang, M. Raymond, S. Rai, D. Werho, D. Roan, B. E. Jr. White and P. J. Tobin: *Impact of titanium addition on film characteristics of HfO_2 gate dielectrics deposited by atomic layer deposition*. Journal of Applied Physics, 98(5):054104, 2005.

[176] Triyoso, D. H., R. Gregory, M. Park, K. Wang and S.I Lee: *Physical and Electrical Properties of Atomic-Layer-Deposited $Hf_xZr_{1-x}O_2$ with TEMAHf, TEMAZr, and Ozone*. Journal of The Electrochemical Society, 155(1):H43–H46, 2008.

[177] Triyoso, D. H., G. Spencer, R. I. Hegde, R. Gregory and C.-D Wang: *Laser annealed $Hf_xZr_{1-x}O_2$ high-k dielectric: impact on morphology, microstructure, and electrical properties*. Applied Physics Letters, 92(11):113501, 2008.

[178] Kim, H., P. C. McIntyre and K. C. Saraswat: *Microstructural evolution of ZrO_2–HfO_2 nanolaminate structures grown by atomic layer deposition*. Journal of Materials Research, 19(2):643–650, 2004.

Figures

2.1 a) Transient current response of b) a polarization hysteresis measurement via triangular field sweeps (inset). The corresponding atomic double-well potential at c) zero field and d) around the coercive field are shown at the bottom. ... 4

2.2 Classification of ferroelectric within the group of (crystalline) dielectrics: Ferroelectric are a subgroup of pyroelectric, which themselves are a subgroup of piezoelectrics. 6

2.3 Sketch of the tetragonal perovskite unit cell of PZT 7

2.4 Second-order transition: (a) Free energy as a function of polarization (ion position) at different temperatures, (b) spontaneous polarization (here, P_0) and (c) susceptibility as function of temperature.[28] 10

2.5 First-order transition: (a) Free energy as a function of polarization (ion position) at different temperatures, (b) spontaneous polarization (here, P_0) and (c) susceptibility as function of temperature.[28] 11

2.6 Different scenarios with polarization hystereses (P-E plots) in the upper and the corresponding current response to a triangular excitation signal (I-E plots) in the lower part of a) – h), respectively. Explanations for possible causes of the curves are given in the text.[48] 14

2.7 Memory hierarchy (left) and classical representatives for the classes in the hierarchy (right). An apparent memory gap between classical "volatile" and "non-volatile" memory existed in 1980.[93] 23

2.8 Evolution of the main stream representatives of the memories classes.[93] .. 23

2.9 Illustration of the extended voltage-time dilemma.[93] 25

2.10 Comparison of DRAM, 1T1C FeRAM and 1T FeRAM cells. 26

2.11 Unit cells of monoclinic, orthorhombic, tetragonal and cubic hafnia polymorphs in (010)-view with lattice parameters taken from Materlik et al.[61]. (Avogadro version 1.1.1 and Avogadro^2 version 0.8.0[148] were used to the unit cell images.) 30

3.1 Metal-insulator-metal stack: a) before and b) after SC1 etch with established short to the bottom electrode.(X: Si, Gd, Sr) 32

3.2 Sketch of the features in a XRR plot for a single layer.[159] 33

3.3 Comparison of XRD geometries: a) Bragg-Brentano geometry vs. b) grazing incidence geometry.[159] 34

3.4 Comparison of different setups for P-E measurements: a) Sawyer-Tower circuit, b) shunt method, c) virtual ground approach.[48] 35

3.5 a) Principle of SF-PFM with sinusoidal small-signal excitation and piezoelectric response. b) Typical amplitude and phase map with corresponding line profiles.[165] 36

3.6 Typical loops for switching spectroscopy: a) piezoelectric response (PR) for e ferroelectric sample consisting of b) amplitude (blue curve; unfolded violet curve equals PR shown in a)) and c) phase. 37

3.7 Band excitation (BE) approach: a) Operational principle of the BE method in SPM. The excitation signal is digitally synthesized to have a predefined amplitude and phase in the given frequency window. The cantilever response is detected and Fourier transformed at each pixel in an image. The ratio of the fast Fourier transform (FFT) of response and excitation signals yields the cantilever response (transfer function). b) Schematic illustration of a SHO corresponding to mass on a spring with damping. Amplitude and phase versus frequency for the SHO model around resonance. SHO fitting yields amplitude, resonance frequency, and Q-factor, that are plotted to yield 2D images, or used as feedback signals. (adapted from [168, 167]) 39

3.8 Comparison of KPFM and cKPFM: a) Applied excitation signals for read and write. KPFM uses a bias sweep to determine the surface potential, whereas a stroboscopic approach (repetition of the bias pulsing sequence with different read voltage levels) is used for cKPFM. b) Surface potential maps from KPFM and cKPFM measurements after charge injection with ± 5 V applied to the scanning tip. c) Line scans across the two regions via KPFM and cKPFM in the left graph and averaged cKPFM curves from the two differently poled areas in the right graph. The area-averaged cKPFM graph also illustrates how the surface potential is obtained from as the x-axis intercept, where the cKPFM signal becomes zero. (adapted from [170]) 41

3.9 Sketch of the RevSTEM approach: a) influence of drift in STEM images, b) measurement of different frames rotated around the image normal; c) uncorrected raw image from a single scan (left) and drift corrected image reconstruction (right). (Figure consists of different images by LeBeau et al.[180]) . 44

3.10 Examples of simulated patterns to show the sensitivity of PACBED to sample polarity ($PbTiO_3$), thickness, tilt and the choice of the semi-convergence angle of the electron beam (all $SrTiO_3$). (Figure consists of different images from [182]) 46

3.11 Linear response theory for fixed voltage excitation (admittance perspective): a) sinusoidal excitation and response signals; b) corresponding phasor/pointer representation.[185] 47

3.12 a) – d) Simulated impedance spectroscopy results for the simple equivalent circuit (inset of b) discussed in the text: Bode (real and imaginary part vs. frequency) and Nyquist plots (locus, i.e. imaginary vs. real part) of impedance and admittance.[185] 48

3.13 Illustration of the FORC approach: a) excitation waveform, resulting b) I-E and c) P-E response and d) reconstructed switching density. ($E_{r,1}$ and $E_{r,n}$ are the first and last reversal field of the FORC measurement respectively.)[187] . . . 51

3.14 Sketch of the measurement procedure to obtain activation energies from the modified harmonic analysis approach used in this work. The steps 1) to 5) are explained in the text above.[70] 53

3.15 Role of the third harmonic in case of a Preisach model hysteresis (most simple case: uniform switching density): a) modelled polarization response vs. time to a sinusoidal excitation for the third third harmonic (dashed grey line) obeying the Preisach model (red solid) and with a phase shift of 180° (blue solid line), b) resulting P-E unpinched and pinched hystereses.[48] 54

4.1 Overview of the impact of different parameters on phase stability as calculated by Materlik et al.[61] using the example of $Hf_{1-x}Zr_xO_2$. ΔE is the difference in free energy between ferroelectric and monoclinic phase originating from the respective contributions.[93] 62

4.2 Example outlining the difficulties in clear phase identification from GIXRD (27 nm thick Gd:HfO_2 film; N_2 anneal at 650 °C for 20 s): a) ferroelectric polarization and small-signal capacitance hysteresis (triangular field sweep with aix-ACCT TF Analyzer 3000) and b) corresponding GIXRD measurement curve and reference patterns for different phases from powder diffraction files.[171, 93] 63

4.3 a) Bright field STEM image showing the film microstructure. The growth direction (G. D.) is indicated by the black arrow. b) HAADF-STEM images acquired from four different grains superimposed with the Hf atom column arrangement projected along the four major zone axes, which are the same for $Pca2_1$, Pbca, and Pbcm phases. The probe semi-convergence angle was 14 mrad for [010] and [110] images, and 20 mrad for [100]and [001] images. c) The crystal structure of five HfO_2 phases projected along four major zone axes. Lattice vectors settings: vector a is red, vector b is green, and vector c is blue.[171] 64

4.4 a) STEM image of the Gd:HfO_2 thin film with a $2 \times 2\,nm^2$ inset showing the projected structure ([110]) corresponding to the experimental PACBED pattern in b). The sample thickness is 10 nm, as determined by PACBED. The growth direction (G.D.) is indicated by the black arrow. Simulated PACBED patterns for the c) polar orthorhombic $Pca2_1$, d) non-polar orthorhombic Pbca, and e) non-polar orthorhombic Pbcm phases. Sample thickness was simulated to match the experiment. Solid and dashed bars indicate the presence or absence of mirror symmetry, respectively.[171] 65

4.5 Crystallization temperature for 10 nm thick FE HfO_2 films with different dopants as obtained from temperature dependent XRD measurements. Data from J. Müller[150] (pure HfO_2, $Si:HfO_2$ and $Y:HfO_2$) as well as measurements performed at Fraunhofer IPMS, Dresden ($La:HfO_2$) or Imec, Leuven and Ghent University ($Al:HfO_2$, $Gd:HfO_2$, $Sr:HfO_2$). 66

4.6 a) GIXRD results of for 30 nm thick $Si:HfO_2$ films with respective reference patterns (see Table 3.2), b) polarization hysteresis and corresponding transient currents for selected compositions and c) Si content dependent evolution of peak position in the transient current curves and sketch of the corresponding bending of the atomic potentials for paraelectric (PE) phase, ferroelectric (FE) phase as well as for the field-induced phase transition into the FE phase (FFE). An ALD cycles ratio of 14:1 results in 5 cat% Si incorporation (XPS). 69

4.7 Wake-up behavior evident from polarization hysteresis in a), c) and corresponding transient currents in b), d) for two characteristic samples: ALD cycle ratio of $HfO_2:SiO_2$ of 28:1 corresponding to an FE sample and a ratio of 20:1 corresponding to a sample at the border between FE and PE as judged by the diffractograms in Fig. 4.6. 70

4.8 GIXRD results of 10 nm thick Sr doped films (anneal: N_2, 800 °C, 20 s) and corresponding reference patterns (powder diffraction files see Table 3.2). This figure plots the data of ref. [141] in logarithmic form. 71

4.9 Selection of 10 nm thick $Sr:HfO_2$ films (N_2 anneal at 800 °C for 20 s): a) Relative permittivity k, remanent polarization P_r, coercive field E_c and switching field E_s over ALD pulse ratio (dashed lines serve as a guide for the eye), b) to d) Large-signal P-E and small-signal C-E (stepwise field sweep via Keithley 4200-SCS) measurements from pure HfO_2 3.4 cat% Sr and 7.9 cat% Sr, respectively. (reproduced from Schenk et al.[141]) 73

4.10 Remanent and saturation polarization P_r and P_{sat} as a function of dopant content for 10 nm thick $Sr:HfO_2$ and $Si:HfO_2$: For Sr, both P_r and P_{sat} decrease simultaneously after a maximum is reached. In contrast, P_{sat} stays still at higher levels for Si even when the maximum in P_r is passed. This is due to a field-induced phase transition that occurs from a tetragonal paraelectric into the orthorhombic ferroelectric phase. The films with highest Sr contents were fabricated using different ALD supercycling schemes and are thus shown in lighter color. 75

4.11 Impact of Si and Sr doping on total energy from ab-initio simulations: While the relative energy reduction of the orthorhombic phase for both Si and Sr doping is favorable and opens a window for a stable FE phase, the strong energy reduction of the tetragonal phase counteract this effect for Si. The tetragonal phase is also favored for Sr doping, but this effect is by far less severe. The arrows indicate the desired situation of reducing the energy difference of the orthorhombic phase and increasing or keeping it constant for the tetragonal phase.[141] 75

4.12 a) GIXRD patterns for different Gd:HfO$_2$ film thicknesses and HfO$_2$ reference patterns; b) relative permittivity k, remanent polarization P_r, and coercive field E_c as a function of Gd:HfO$_2$ thickness; c) evolution of P_r with film thickness compared to other dopants. (reproduced from Hoffmann et al.[172]) . . . 77

4.13 a) Grazing incidence X-ray diffractograms for Gd:HfO$_2$ samples and reference patterns are given for the paraelectric monoclinic, tetragonal and cubic phases as well as for the ferroelectric orthorhombic phase. b) Comparison of the polarization hystereses resulting from different annealing conditions after 10^4 field cycles.[254] 78

4.14 a) Dark field TEM shows Gd:HfO$_2$ grain sizes of about 27 nm. b) Bright field TEM of the BE interface. The arrow indicates a Gd:HfO$_2$ grain boundary which coincides with a TiN boundary.[172] 79

4.15 Example of Gd:HfO2 (3.4 cat%): a) GIXRD measurement and reference patterns and b) relative permittivity k and remanent polarization P_r for different TE/BE configurations and annealing conditions.[172] 80

4.16 ToF-SIMS results for a) TiN/TiN, b) TiN/TaN, and c) TaN/TaN samples annealed at 650 °C for 20 s. Bars on the y-axis indicate a factor of 10 in intensity ratio.[172] The SIMS measurements were performed by Maximilian Drescher at Fraunhofer IPMS, Dresden. 81

4.17 Average change in the total energy difference ΔE_{tot} between the various phases of HfO$_2$ compared to the monoclinic phase as a function of the oxygen vacancy concentration.[172] 82

4.18 GIXRD scans for different rotations around the plane normal (angle Φ) in fine steps of 5° for $\Phi = 0°...60°$ and coarse steps of $\Phi = 15°$ for 60°... 80°. No significant in-plane texture is observed. This sample (Si:HfO$_2$, ALD cycle ratio of 26:1, thickness of 30 nm) exhibits a so-called fiber texture. 82

4.19 Rietveld refinement of 2Θ-Θ-scans for different Ψ (Si:HfO$_2$, ALD cycle ratio of 26:1, thickness of 30 nm): Phase fractions of 25 mol% monoclinic, 40 mol% orthorhombic HfO$_2$ and 35 mol% TiN were obtained. The orthorhombic phase was approximated with a tetragonal structure to reduce the amount of fit parameters in the refinement. A polynomial background fit and a spherical harmonic texture model was used. Moreover, underground peaks (UP) were introduced to account for substrate-related and other artifacts such as double diffractions or fluorescence phenomena. For the sake of convenience, the location of main peaks of all three phases are shown in the diagram in a simplified manner. The * symbol indicates the presence of multiple non-equivalent families of lattice plane that cause reflexes at different locations around the position marked by the respective arrow. The Rietveld refinement was performed by Christopher M. Fancher (CNMS, Oak Ridge National Laboratories). 84

4.20 Texture assessment: a) Sketch of the angles in the used Bragg-Brentano geometry; b) Example pole figures reconstructed from the texture spherical harmonic model show only a weak mixed texture of up to 3 MRD (multiples of random distribution). Within the uncertainties of the refinement, the results do not justify to conclude that a significant preferential orientation is present in the HfO$_2$ or TiN and the orientation is rather random. 86

4.21 Characteristic plot of the $\sin^2(\Psi)$ method[258]: Only the (111) peak of the FE phase exhibits a significant shift between in- and out-of-plane orientation. The difference in lattice plane spacing is found to be around 1 %. If elastic properties were known, the residual stress could be calculated from the plot. . 87

4.22 Sketch of incipient ferroelectricity: Temperature dependence of dielectric behavior anticipated from Curie-Weiss law (Eq. 2.12) and deviation from that case for incipient ferroelectricity (modeled as described by Barrett et al.[263]) 88

4.23 Qualitative model of the phase transitions of HfO$_2$ influenced by the most important process parameters.[172] 90

5.1 Schematic of the experimental observations for the three phenomena condensed in the term "electric field cycling behavior".[187] 91

5.2 a) Preisach plane and the switching fields of the respective bistable units in different sections of the plane. b) Sketch of the contour-plots to be observed for different scenarios from Fig. 2.4. The ellipses show the location of a maximum in the switching density. ($E_{sw+/-}$ denote the field necessary to switch to positive/negative polarization state. E'_{sw} and E'_{bias} are transforms of alternative coordinates for the Preisach plane that represent the mean absolute value of the switching field and the internal bias field, respectively).[48] 92

5.3 Wake-up effect: The constricted hysteresis of a pristine sample opens after field cycling and two distinct current peaks merge into one single peak. The corresponding P-E hystereses are shown as inset with the E- and P-axes scaled from $-4\,\text{MV/cm}$ to $4\,\text{MV/cm}$ and from $-30\,\mu\text{C/cm}^2$ to $30\,\mu\text{C/cm}^2$, respectively. The cycling sequence is sketched next to the graph.[48] 93

5.4 Extraction of activation energies via harmonic analysis: a) Example of the evolution of amplitude (normalized to the respective initial value) and b) phase against the number of fatigue cycles for $f = 10\,\text{kHz}$ and $T = 171\,\text{K}$. Jumps in the phase of the seventh and ninth harmonics occur subsequently and are accompanied by a minimum in the amplitude. c) Arrhenius plot of the cycling time t_i at which the phase jumps of the respective i-th harmonic was observed. Data points were extracted from a) as shown and similarly for the other temperatures between $136\,\text{K}$ and $298\,\text{K}$.[48] 93

5.5 Wake-up for $4\,\text{MV/cm}$ amplitude: Experimental switching density determined for a) a pristine capacitor, b) after 100, and c) after 10^4 rectangular switching cycles. The insets show the respective polarization hystereses (extracted from the last sweep of the corresponding FORC measurement) for comparison. The solid line of each inset is the hysteresis for the accompanying switching density plot, whereas the dashed lines are the respective hystereses of the two other cycling stages for comparison.[187] 94

5.6 Problem of internal bias fields for low-power-operation of ferroelectric memories: a) Evolution of $2P_r$ for different field amplitudes as a function of the number of cycles and b) switching density of the initial sample obtained with a FORC amplitude of $4.5\,\text{MV/cm}$. The triangular white frames represent the accessible range of domains for the respective lower field amplitudes. Internal bias fields decrease the amount of domains switchable in both polarization directions.[187] 95

5.7 Arrhenius plot for the fatigue at different frequencies for $4\,\text{MV/cm}$ field cycling amplitude: Number of cycles $n_{P_r,max}$ at which the maximum remanent polarization was observed vs. inverse temperature.[48] 96

5.8 Fatigue for $3\,\text{MV/cm}$ amplitude at $10\,\text{kHz}$: a) Evolution of $2P_r$ and the peak height(s) of the switching density ρ_{max} with the number of switching cycles and corresponding switching density plots obtained b) after 10^4 cycles, i.e. after wake-up and c) after 10^6 cycles, i.e., in a fatigued state. The solid line of each inset is the hysteresis for the accompanying switching density plot, whereas the dashed lines are the respective hystereses of the two other cycling stages for comparison.[187] 98

5.9 a) Pinching of the P-E hysteresis and b) Split-up of the corresponding currents peaks in I-E graphs with increasing number of cycles with an electric field of $2\,\text{MV/cm}$.[48] 99

5.10 a) Current peak split-ups occur directly at the values of the cycling field amplitude (constant no. of cycles). b) Split-up of multiple peaks by subsequent field cycling with different field amplitudes in descending order. The split-up at 1.45 MV/cm has already vanished in the first quadrant (red line). All experiments were performed with $f = 1000\,\text{Hz}$. The corresponding P-E hystereses are shown as insets with the E- and P-axes scaled from $-4\,\text{MV/cm}$ to $4\,\text{MV/cm}$ and from $-30\,\text{µC/cm}^2$ to $30\,\text{µC/cm}^2$, respectively.[187] 99

5.11 Split-up/merging for 4 MV/cm amplitude after wake-up treatment (see Fig. 5.5 c)): Experimental switching density determined for a) a capacitor subjected to 10^6 cycles of 2 MV/cm (split-up) amplitude, b) the same capacitor after 10^4 merging cycles of 4 MV/cm amplitude, and c) a capacitor subjected to a double split-up by subsequent cycling at 2.5 MV/cm and 1.5 MV/cm for 10^6 cycles each. Note the different pseudocolor-scale of (c). The solid line of each inset is the hysteresis for the accompanying switching density plot, whereas the dashed lines are the respective hystereses of the two other cycling stages for comparison.[187] 101

5.12 Evolution of the polarization state during the pulse train of the split-up experiments. See text for a detailed explanation of the manifestation of the split-up effect as a local imprint/internal biasing mechanism with a) – c) one and d) two subcycling field(s).[187] 103

5.13 Evolution of the hysteresis shape during wake-up for 100 Hz and 10000 Hz. For higher frequencies, lower P_r values are observed and the hysteresis edges are less steep. 105

5.14 SF-PFM results on bare oxide for a 27 nm thick Gd:HfO$_2$ sample a) in a preliminary experiment and b) before BEPS and cKPFM and with the same tip in the same spot as these subsequent measurements. c) The frequency sweep prior to the series of SF-PFM scans, BEPS and cKPFM measurements: A good signal-to-noise ratio is observed. Frequency sweeps after the measurement sequence usually exhibited amplitudes that were at least a factor of ten lower. 110

5.15 Off-field BEPS on bare HfO_2: Evolution of amplitude A_0, phase angle φ, contact resonance frequency ω_0 and quality factor Q during the sketched bias voltage sweep as obtained from the SHO fit of the response in frequency domain (after Fourier transform). At 0 V only one phase and varying amplitudes are observed. The samples has a preferential polarization orientation across the whole probes area. At 8.75 V, part of the domains are switched to opposite polarization (phase changed be 180°, now red color). White pixels denote that no successfull SHO fit was possible due to the low signal around the switching field. At 14 V nearly the whole area is switched to the opposite state. Backswitching to the initial state (blue color in the phase diagrams) starts at around -1.75 V, i.e. at lower absolute voltage values than the opposite switching. At -14 V, the whole sample area is in the initial polarization state. A and B denote examples of the two types of differently biased regions found in the sample. 112

5.16 Electrostatic contributions to the piezoresponse (common symbols in publications: PR, PFM, $A \cdot \sin(P)$): a) sketch of the interaction of purely electrostatic and purely piezoelectric signals[166], b) comparison of on- and off-field loops for PZT and amorphous HfO_2[286] and c) averaged on- and off-field loops extracted from BEPS measurements in Fig. 5.15. 113

5.17 Comparison of cKPFM results for the 27 nm thick $Gd:HfO_2$ film of this work to what Balke et al. showed for a ferroelectric PZT film and an amorphous and therefore non-ferroelectric HfO_2 film.[286] 115

5.18 On- and off-field loops of the piezoresponse $PR = A_0 \cdot \cos(\varphi + \Delta\varphi)$ (phase offset $\Delta\varphi$ chosen to maximize loop area) from BEPS averaged over the 50×50 pixels in Fig. 5.15 compared to the results for PZT and amorphous HfO_2 by Balke et al.[286] 116

5.19 a) Hystereses of polarization P (dotted lines) and relative permittivity k (solid lines) obtained from small-signal capacitance measurements for the three different regimes: pristine (0 cycles), during wake-up (1000 cycles) and during fatigue (215000 cycles); b) the trend of remnant polarization P_r and minimum relative permittivity $k_{r,min}$ vs. number of switching cycles; c) a bright field STEM image showing the whole film stack and HfO_2 grains that span the whole film thickness.[185] 118

5.20 Impedance spectroscopy: a) logarithmic Bode plots of impedance Z and admittance Y (inset); b) Nyquist plots of impedance and admittance in both full scale and magnified sections to show the high quality of the fit throughout the whole frequency spectrum. Blue, green and orange color represent the capacitors in pristine state, after 1000 cycles and after 215000 cycles at 1 kHz and 8.5 V, respectively[185] 121

5.21 Impedance spectroscopy: Film stack with corresponding equivalent circuit. Fit parameters can be found in Table 5.3[185] 121

5.22 Leakage current characteristics for capacitors in a pristine state (blue), after 1000 (toward the end of wake-up, orange) and after 215000 cycles (during fatigue, green). The arrows exemplarily indicate by what factor the leakage current of the pristine sample at 1 V multiplied during wake-up and fatigue.[185] 122

5.23 Examples of HAADF-STEM micrographs: The interfacial HfO_2 layer undergoes complex changes in interfacial strain and phase presence with field cycling, including some reduction of the presence of the tetragonal phase in the cycled (woken-up and fatigued) samples compared to the pristine capacitor. Lattice parameter maps from these regions further highlight the complex interface environments and assist in visualizing strain and phase presence (colorscale ranges set manually to better visualize details).[185] 124

5.24 EELS image (left) of the woken-up sample with white tick marks showing layers of the stack, and with colored rectangles 1 – 3 indicating regions in the Ti adhesion layer where the signal from the spectrum image was integrated and plotted (right). The spectra are very similar at each location, indicating relative uniformity of oxygen content. The Ti L-edge and O K-edge onsets are marked, and occur near their known values of 456 eV and 532 eV respectively.[185] 125

5.25 Effect of non-uniform defect-rich, charged layers adjacent to the electrodes: a) Capacitor stack with four different cases of positive charge only at the bottom electrode (I), at both electrodes (II), only at the top electrode (III) and at none of the electrodes (IV). b) Manifestation of the aforementioned scenarios in the macroscopically measures P-E hysteresis. [185] 126

Tables

2.1 Comparison of mainstream memories and FRAM. 25
3.1 Process parameters used to fabricate hafnia based MIM capacitors with different dopant materials. (RT = room temperature) 32
3.2 Numbers of the PDF reference data (International Centre for Diffraction Data, PDF-4+ 2014, database version 4.1403) used for phase analysis. 34
3.3 Comparison of electron diffraction methods: conventional electron diffraction, convergent beam electron diffraction (CBED) and position averaged convergent beam electron diffractions (PACBED). 45
4.1 Ionic radii in pm of Hf and different dopant elements in different coordination numbers in a crystalline lattice.[227, 228] The lines in bold font indicate the relevant coordination numbers in the respective phases indicated by the superscripts: M – monoclinic $P2_1/c$, O – orthorhombic $Pca2_1$, T – tetragonal $P4_2/nmc$, C – cubic $Fm3m$ 58
5.1 Comparison of activation energies for wake-up, fatigue, and remerging after split-up.[70] 104
5.2 Relative permittivities from literature for the ferroelectric orthorhombic $Pca2_1$[61, 132] and the paraelectric monoclinic $P2_1/c$[61, 132, 133, 134, 135], tetragonal $P4_2/nmc$[61, 133, 135], and cubic $Fm3m$[61, 132, 133, 134, 135] phases.[185] . 118
5.3 Impedance spectroscopy: Model parameters of the equivalent circuits sketched in Fig. 5.21 used to fit the data presented in Fig. 5.20. Relative standard deviations of all fit parameters are 10^{-4} or below.[185] 122
5.4 Tabulated monoclinic (M) and orthorhombic (O) phase fractions measured from the pristine, woken-up, and fatigued samples together with corresponding remanent/maximum polarization P_r/P_{max} values.[185] 123

Abbreviations

Abbreviation	Explanation
1T	one transistor
1T1C	one transistor one capacitor
2T2C	two transistor two capacitor
ALD	atomic layer deposition
AC	alternating current
AFE	antiferroelectric
BE	bottom electrode
BEPS	band-excitation point spectroscopy
BF	bright field
BL	bit-line
cat%	catinonic percent, ratio of dopant/(dopant + Hf) atoms
CBED	convergent-beam electron diffraction
cKPFM	contact-mode Kelvin-probe force microscopy
CMOS	complementary metal-oxide-semiconductor
CNMS	Center for Nanophase Materials Sciences
CPE	constant-phase element
CPU	central processing unit
CSD	chemical solution deposition (sol-gel method)
CVD	chemical vapor deposition
DC	direct current
DF	dark field
DLCC	dynamic leakage current compensation[50, 51]
DRAM	dynamic random access memory
EDS	energy dispersive X-ray spectroscopy, also called EDX
EDX	energy dispersive X-ray spectroscopy,, also called EDS
EELS	electron energy loss spectroscopy
FE	ferroelectric
FeFET	ferroelectric field effect transistor
FERAM	ferroelectric random access memory
FeRAM	ferroelectric random access memory
FET	field effect transistor
FFE	field-induced ferroelectricity
FFT	fast Fourier transform
FORC	first-order reversal curves
FRAM	ferroelectric random access memory
f.u.	formula unit (of a chemical compound)
GATR-FTIR	grazing angle total reflection Fourier transform infrared spectroscopy

Abbreviation	Explanation
G.D.	growth direction
GIXRD	grazing incidence X-ray diffraction
GND	ground
HAADF	high-angle annular dark field
HR-TEM	high-resolution transmission electron microscopy
ICDD	International Centre for Diffraction Data
IS	impedance spectroscopy
KPFM	Kelvin-probe force microscopy
MIM	metal-insulator-metal (stack for a capacitor)
NV	non-volatile
PACBED	position-averaged convergent-beam electron diffraction
PDF	powder diffraction file
PE-ALD	plasma enhanced atomic layer deposition
PFM	piezoresponse force microscopy
PL	plate-line
PZT	lead zirconate titanate, $Pb(Ti,Zr)O_3$
PVD	physical vapor deposition
RAM	random access memory
RevSTEM	revolving scanning transmission electron microscopy
SC1	standard clean 1
SF-PFM	single-frequency piezoresponse force microscopy
SHO	simple harmonic oscillator
SPM	scanning probe microscopy
SSD	solid-state drive
STEM	scanning transmission electron microscopy
TE	top electrode
TEM	transmission electron microscopy
ToF-SIMS	time-of-flight secondary ion mass spectrometry
WL	word-line
XPS	X-ray photo electron spectroscopy
XRD	X-ray diffraction
XRR	X-ray reflectometry

Symbols

Symbol	Unit	Explanation
2θ	°	angle between X-ray tube and detector
ΔE	arb. unit/f.u.	energy difference per formula unit (f.u.) between the ferroelectric and the monoclinic phase
ε	$F \cdot m^{-1}$	permittivity
ε_0	$F \cdot m^{-1}$	electric constant, $8.85... \cdot 10^{-12}\ F \cdot m^{-1}$
ε_r	1	relative permittivity, also denoted with k
θ	°	angle between of X-ray tube and fixed chuck
λ	m	wavelength
π	1	Archimedes constant, ratio of circumference to diameter of a circle, 3.14...
ρ	$kg \cdot m^{-3}$	(mass) density
ρ^-	$C \cdot V^{-2}$	switching (density) obtained from a FORC measurement with decreasing reversal fields E_r
ρ^-_{FORC}	$C \cdot V^{-2}$	switching (density) obtained from a FORC measurement with decreasing reversal fields E_r
ρ_{max}	$C \cdot V^{-2}$	maximum in switching (density)
Φ	°	angle of rotation around the surface normal of a sample in XRD
ϕ	°	phase angle
φ	°	angle in the plane of the sample surface
φ	°	phase angle
χ	1	electric susceptibility
Ψ	1	wave function
ω	s^{-1}	angular frequency
ω	°	angle between fixed X-ray tube and chuck
ω_0	s^{-1}	(angular) resonance frequency
A	m^2	capacitor area
A	arb. unit	amplitude of the measured piezoresponse
A	arb. unit	amplitude of sample's piezoresponse
a	m	lattice constant along [100] direction
a_0	$C \cdot m^{-2}$	coefficient of the second-order polynomial fit for the switching density in the FORC measurements E_r
$a_{1/2}$	$C \cdot V^{-1} \cdot m^{-1}$	coefficient of the second-order polynomial fit for the switching density in the FORC measurements E_r
$a_{3/4/5}$	$C \cdot V^{-2}$	coefficient of the second-order polynomial fit for the switching density in the FORC measurements E_r
B	T	magnetic flux density
b	m	lattice constant along [010] direction

List of Symbols

Symbol	Unit	Explanation
C	F	capacitance
$C_{1/3}$	Ω	capacitance in the first/third RC-like element in the equivalent circuit for impedance spectroscopy
C_{FE}	F	capacitance of the ferroelectric layer
C_{IF}	F	capacitance of the dielectric interface layer
C_p	F	parallel capacitance
C_{ref}	F	reference capacitance
c	m	lattice constant along [001] direction
$c_{2/4/6}$	$m \cdot F^{-1}$	coefficients of the power series expansion in Landau theory
$cKPFM$	arb. unit	signed measured piezoresponse as used in cKPFM
D	$C \cdot m^{-2}$	(electric) displacement field
D_0	$C \cdot m^{-2}$	(electric) displacement field of vacuum
d	m	film thickness, separation of capacitor plates
d_{33}	m/V	coefficient of the piezoelectric tensor describing the strain along the polar [001] direction of HfO_2
d_{eff}	m/V	effective piezoelectric constant in field direction
d_{FE}	m	thickness of the ferroelectric layer
d_{IF}	m	thickness of the dielectric interface layer
E	$V \cdot m^{-1}$	electric field
E_{BD}	$V \cdot m^{-1}$	(dielectric) breakdown field
E_{bs}	$V \cdot m^{-1}$	(electric) backswitching field (i.e. to negative polarization state), also denoted as E_{sw-}
E_{bias}	$V \cdot m^{-1}$	(electric) bias field
E_c	$V \cdot m^{-1}$	coercive (electric) field
$E_{c+/-}$	$V \cdot m^{-1}$	positive/negative coercive (electric) field
E_{depol}	$V \cdot m^{-1}$	(electric) depolarization field
E_{gap}	J	band gap energy
E_{FE}	$V \cdot m^{-1}$	(electric) field across the ferroelectric layer
E_{IF}	$V \cdot m^{-1}$	(electric) field across the dielectric interface layer
E_r	$V \cdot m^{-1}$	(electric) reversal field in the FORC measurement
$E_{r,1}$	$V \cdot m^{-1}$	first (electric) reversal field in the FORC measurement
$E_{r,i}$	$V \cdot m^{-1}$	i-th (electric) reversal field in the FORC measurement
$E_{r,i-1}$	$V \cdot m^{-1}$	$i-1$-th (electric) reversal field in the FORC measurement
$E_{r,n}$	$V \cdot m^{-1}$	last (electric) reversal field in the FORC measurement
E_s	$V \cdot m^{-1}$	(electric) switching field (i.e. to positive polarization state), also denoted as E_{sw+}

Symbol	Unit	Explanation
E_{sw+}	$V \cdot m^{-1}$	(electric) switching field (i.e. to positive polarization state), also denoted as E_s
E_{sw-}	$V \cdot m^{-1}$	(electric) backswitching field (i.e. to negative polarization state), also denoted as E_{bs}
E_{sat}	$V \cdot m^{-1}$	(electric) saturation field
$E_{sub/sub1/sub2}$	$V \cdot m^{-1}$	subcycling (electric) field, i.e. a field of non-saturating amplitude
$E_{sub1/2}$	$V \cdot m^{-1}$	first/second subcycling (electric) field, i.e. a field of non-saturating amplitude
E_{tot}	J/f.u.	total energy (internal energy per formula unit f.u.)
e	1	Euler constant, 2.72...
$F_{Coulomb}$	N	Coulomb force
f	Hz	frequency
G	J	Gibbs energy
g	$J \cdot m^{-3}$	Gibbs energy per volume
H	$A \cdot m^{-1}$	magnetic field strength
I	A	electrical current
I_0	V	amplitude of the sinusoidal current response for impedance spectroscopy
I_{AC}	A	alternating current
I_C	A	dielectric current
I_D	A	electrical displacement current
I_{fe}	A	ferroelectric current
I_R	A	(Ohmic) leakage current
i	1	running index
j	$A \cdot m^{-2}$	current density
j^-_{FORC}	$A \cdot m^{-2}$	current density obtained from a FORC measurement with decreasing reversal fields E_r
k	1	relative permittivity, also denoted with ϵ_r
k_{FE}	1	relative permittivity of the ferroelectric layer
k_{IF}	1	relative permittivity of the dielectric interface layer
n	1	exponent of the constant-phase element (CPE), sometimes called "pseudo-capacitance"
P	$C \cdot m^{-2}$	polarization (density)
P_{FE}	$C \cdot m^{-2}$	polarization (density) of the ferroelectric layer
P^-_{FORC}	$C \cdot m^{-2}$	polarization (density) obtained from a FORC measurement with decreasing reversal fields E_r
P_r	$C \cdot m^{-2}$	remanent polarization (density)
$P_{r+/-}$	$C \cdot m^{-2}$	positive/negative remanent polarization (density)

List of Symbols

Symbol	Unit	Explanation
P_s	$C \cdot m^{-2}$	spontaneous polarization (density)
PFM	arb. unit	signed measured piezoresponse, sometimes called "unfolded" piezoresponse
PR	arb. unit	signed measured piezoresponse, sometimes called "unfolded" piezoresponse
p	$C \cdot m^{-2}$	dipole moment
R	Ω	(electrical) resistance
$R_{1/2/3}$	Ω	(electrical) parallel resistance as used in the second, third and forth RC-like element in the equivalent circuit for impedance spectroscopy
R_4	Ω	(electrical) series resistance as used in the equivalent circuit for impedance spectroscopy
R_{ref}	Ω	(electrical) reference resistance
R_p	Ω	(electrical) parallel resistance
R_s	Ω	(electrical) series resistance
Q	C	charge
Q	Ω	impedance of a constant-phase element (CPE)
Q	1	quality factor of the simple harmonic oscillator model
Q_2	Ω	impedance of the constant-phase element (CPE) in the second RC-like element in the equivalent circuit for impedance spectroscopy
Q_{FE}	C	charge of the ferroelectric layer
Q_{IF}	C	charge of the dielectric interface layer
Q_{ion}	C	charge of a single ion
T	K	temperature
T	$F \cdot s^n$	prefactor of the constant-phase element (CPE), sometimes called "pseudo-capacitance"
T_0	K	phase transition temperature
T_C	K	Curie temperature
T_{cryst}	K	crystallization temperature
t	s	time
$t_{1/2/.../9}$	s	time until the phase jump of first/second/.../ninth harmonic of the polarization response in harmonic analysis
V	V	voltage
V_0	V	amplitude of the sinusoidal voltage excitation for impedance spectroscopy
V_{AC}	V	voltage of an alternating current signal
V_{ac}	V	voltage of an alternating current signal

Symbol	Unit	Explanation
V_{bias}	V	voltage of a direct current signal, bias voltage, also denoted as V_{DC}
V_C	V	voltage across the reference capacitance
V_{DC}	V	voltage of a direct current signal, bias voltage also denoted as V_{bias}
V_{dc}	V	voltage of an alternating current signal
V_{ext}	V	external voltage excitation
V_{FE}	V	voltage across the ferroelectric layer
V_R	V	voltage across the reference resistance
V_{read}	V	read (DC) voltage level in cKPFM
V_{write}	V	write (DC) voltage level in cKPFM
X	Ω	(electrical) reactance
x	m	displacement, ion position
Y	S	(electrical) admittance
Z	1	atomic number (typically used in "Z contrast" of TEM)
Z	Ω	(electrical) impedance

Curriculum Vitae

Tony Schenk

Date of birth	January 11th 1987
Place of birth	Großenhain, Germany
Nationality	German
E-mail address	tony.schenk@web.de

ACADEMIC CAREER

Scientist/PhD student
Jan. 2013 – present
NaMLab gGmbH/TU Dresden: Research on ferroelectric HfO_2

Master studies (M.Eng)
Mar. 2011 – Dec. 2012
Nano and Surface Technology, UAS Zwickau, master thesis "Plasma Enhanced Atomic Layer Deposition and Characterization of Ferroelectric Aluminum Doped Hafnium Oxide"

Bachelor studies (B.Eng)
Dec. 2010 – Mar. 2011
Microtechnology (dual), UAS Zwickau in cooperation with Infineon Technologies Dresden bachelor thesis: "Application of Current Models and Design of Experiments (DoE) to Chemical-Mechanical Polishing of Tungsten and Verification of the Results by a Pad Evaluation"

TEACHING EXPERIENCE

Lecturer
Sept. 2010/2012 – 2016
UAS Zwickau, Germany: Preparatory courses in Physics for freshmen of different faculties

Tutor for "Memory Technology"
Oct. 2013 – Mar. 2015,
NaMLab gGmbH/TU Dresden, Germany: Exercises for master students, advisor of laboratory courses

Tutor for "Physical Chemistry" Oct. 2009 – June 2010, Oct. 2011 – Mar. 2012	UAS Zwickau, Germany: Exercises for second through fourth term students

MISCELLANEOUS

Research stay Nov. 2016	Oak Ridge National Laboratories, Center for Nanophase Materials Sciences (CNMS), scanning probe microscopy
Research stay Apr. – May 2015	Oak Ridge National Laboratories, Center for Nanophase Materials Sciences (CNMS), scanning probe microscopy
Award for master thesis 2014	Category "Engineering", Mentor e.V.
Scholar/Trainee Aug. – Sept. 2011	**Taiwan Summer Institute Program (DAAD)**, NTHU Hsinchu, Taiwan: MATLAB programming for spectral reflectometry
Scholar Sept. 2011 – Aug. 2012	**Deutschlandstipendium**
Trainee Sept. 2006 – Mar. 2011	Infineon Technologies Dresden GmbH, Germany: Dept. Chemical-Mechanical Polishing, process optimization and evaluation of consumables

List of Scientific Publications

Conference Talks

1) **T. Schenk**, M. Hoffmann, J. Ocker, M. Pešić, E. D. Grimley, X. Sang, J. M. LeBeau, T. Mikolajick, U. Schroeder, Internal Bias Fields in Ferroelectric HfO2 Thin Films and their Structural Origins, *Joint IEEE International Symposium on the Applications of Ferroelectrics, European Conference on Applications of Polar Dielectrics & Workshop on Piezoresponse Force Microscopy (ISAF/ECAPD/PFM)*, Darmstadt, 2016.
2) **T. Schenk**, M. Hoffmann, C. Richter, M. Pešić, S. Mueller, S. Slesazeck, U. Schroeder, T. Mikolajick, D. Pohl J. Mueller, P. Polakowski, R. Materlik, A. Kersch X. Sang, E. D. Grimley, J. M. LeBeau, Doped Hafnium Oxide for Ferroelectric Memories, *MRS Fall Meeting, Boston*, 2015. (invited)
3) **T. Schenk**, U. Schroeder, M. Hoffmann, M. Pešić, M. Popovici, Y. V. Pershin, T. Mikolajick, Insights into Analyzing the Electric Field Cycling Behavior of Ferroelectric Doped Hafnium Oxide, *European Conference on Application of Polar Dielectrics (ECAPD)*, Vilnius, 2014.
4) **T. Schenk**, S. Mueller, U. Schroeder, R. Materlik, A. Kersch, M. Popovici, C. Adelmann, S. Van Elshocht, T. Mikolajick, Strontium Doped Hafnium Oxide Thin Films: Wide Process Window for Ferroelectric Memories, *IEEE European Solid-State Device Research Conference (ESSDERC)*, Bucharest, 2013.

Articles in Conference Proceedings as First Author

1) **T. Schenk**, S. Mueller, U. Schroeder, R. Materlik, A. Kersch, M. Popovici, C. Adelmann, S. Van Elshocht, T. Mikolajick, Strontium Doped Hafnium Oxide Thin Films: Wide Process Window for Ferroelectric Memories, *Proceedings of the European Solid-State Device Research Conference (ESSDERC)*, 2013, pp. 260-263.

Journal Articles as First Author

1) E. D. Grimley, **T. Schenk**, X. Sang, M. Pešić, U. Schroeder, T. Mikolajick, J. M. LeBeau, Structural Changes Underlying Field-Cycling Phenomena in Ferroelectric HfO_2 Thin Films, Adv. Electron. Mater.,2, 9, 1600173 (2016).
2) **T. Schenk**, M. Hoffmann, J. Ocker, M. Pešić, T. Mikolajick, U. Schroeder, Complex Internal Bias Fields in Ferroelectric Hafnium Oxide, *ACS Appl. Mater. Interfaces*, 7, 36, 20224-20233 (2015).
3) **T. Schenk**, U. Schroeder, T. Mikolajick, Dynamic Leakage Current Compensation Revisited, *IEEE Trans. Ultrason., Ferroelectr., Freq. Control*, 62, 3, 596-599 (2015).

4) **T. Schenk**, E. Yurchuk, S. Mueller, U. Schroeder, S. Starschich, U. Böttger, T. Mikolajick, About the deformation of ferroelectric hystereses, *Appl. Phys. Rev.*, 1, 041103 (2014).
5) **T. Schenk**, U. Schroeder, Pešić, M. Popovici, Y. V. Pershin, T. Mikolajick, Electric Field Cycling Behavior of Ferroelectric Hafnium Oxide, *ACS Appl. Mater. Interfaces*, 6, 22, 19744-19751 (2014).

Journal Articles as Co-author

1) U. Schroeder, M. Pešić, **T. Schenk**, H. Mulaosmanovic, S. Slesazeck, J. Ocker, C. Richter, E. Yurchuk, K. Khullar, J. Müller, P. Polakowski, E. D. Grimley, J. M. LeBeau, S. Flachowsky, S. Jansen, S. Kolodinski, R. van Bentum, A. Kersch, C. Künneth, T. Mikolajick, Impact of field cycling on HfO_2 based non-volatile memory devices, *Proceedings of the European Solid-State Device Research Conference (ESSDERC)*, 2016, pp. 364-368. (invited)
2) F. P. G. Fengler, M. Pešić, S. Starschich, T. Schneller, U. Böttger, **T. Schenk**, M. H. Park, T. Mikolajick, U. Schroeder, Comparison of hafnia and PZT based ferroelectrics for future non-volatile FRAM applications, *Proceedings of the European Solid-State Device Research Conference (ESSDERC)*, 2016, pp. 369-372.
3) M. Pešić, F. P. G. Fengler, L. Larcher, A. Padovani, **T. Schenk**, E. D. Grimley, X. Sang, J. M. LeBeau, S. Slesazeck, U. Schroeder, T. Mikolajick, Physical Mechanisms Behind the Field-Cycling Behavior of HfO_2 Based Ferroelectric Capacitors, *Adv. Funct. Mater.*, 26, 25, 4601-4612 (2016).
4) D. K. Simon, D. Tröger, **T. Schenk**, I. Dirnstorfer, F. P. G. Fengler, P. M. Jordan, A. Krause, T. Mikolajick,; Comparative study of ITO and TiN fabricated by low-temperature RF biased sputtering, *J. Vac. Sci. Technol. A*, 34, 2, 021503 (2016).
5) M. Pešić, S. Slesazeck, **T. Schenk**, U. Schroeder, T. Mikolajick, Impact of charge trapping on the ferroelectric switching behavior of doped HfO_2, *Phys. Status Solidi A*, 213, 2, 270-273 (2015).
6) Y. Guan, D. Zhou, J. Xu, X. Liu, F. Cao, X. Dong, J. Müller, **T. Schenk**, U. Schroeder, The Rayleigh law in silicon doped hafnium oxide ferroelectric thin films, *Phys. Status Solidi RRL*, 9, 10, 589-593 (2015).
7) M. Hoffmann, U. Schroeder, **T. Schenk**, T. Shimizu, H. Funakubo, O. Sakata, D. Pohl, M. Drescher, C. Adelmann, R. Materlik, A. Kersch, T. Mikolajick, Stabilizing the ferroelectric phase in doped hafnium oxide, *J. Appl. Phys.*, 118, 7, 072006 (2015).
8) D. Zhou, Y. Guan, M. M. Vopson, J. Xu, H. Liang, F. Cao, X. Dong, J. Mueller, **T. Schenk**, U. Schroeder, Electric field and temperature scaling of polarization reversal in silicon doped hafnium oxide ferroelectric thin films, *Acta Mater.*, 99, 240-246 (2015).
9) X. Sang, E. D. Grimley, **T. Schenk**, U. Schroeder, J. M. LeBeau, On the structural origins of ferroelectricity in HfO_2 thin films, *Appl. Phys. Lett.*, 106, 162905 (2015).

10) M. Hoffmann, **T. Schenk**, I. Kulemanov, C. Adelmann, M. Popovici, U. Schroeder, T. Mikolajick, Low Temperature Compatible Hafnium Oxide Based Ferroelectrics, *Ferroelectrics*, 480, 1, 16-23 (2015).

11) D. Martin, J. Müller, **T. Schenk**, T. M. Arruda, A. Kumar, E. Strelcov, E. Yurchuk, S. Müller, D. Pohl, U. Schröder, S. V. Kalinin, T. Mikolajick, Ferroelectricity in Si-doped HfO_2 Revealed: A Binary Lead-free Ferroelectric, *Adv. Mater.*, 26, 48, 8197-8202 (2014).

12) T. Mikolajick, S. Müller, **T. Schenk**, E. Yurchuk, S. Slesazeck, U. Schröder, S. Flachowsky, R. van Bentum, S. Kolodinski, P. Polakowski, J. Müller, Doped Hafnium Oxide—An Enabler for Ferroelectric Field Effect Transistors, *Adv. Sci. Technol.*, 95, 136-145 (2014).

13) C. Richter, **T. Schenk**, U. Schroeder, T. Mikolajick, Film properties of low temperature HfO_2 grown with H_2O, O_3, or remote O_2-plasma, *J. Vac. Sci. Technol. A*, 32, 01A117 (2014).

14) S. Clima, D. J. Wouters, C. Adelmann, **T. Schenk**, U. Schroeder M. Jurczak, G. Pourtois, Identification of the ferroelectric switching process and dopant-dependent switching properties in orthorhombic HfO_2: A first principles insight, *Appl. Phys. Lett.*, 104, 092906 (2014).

15) U. Schroeder, E. Yurchuk, J. Müller, D. Martin, **T. Schenk**, P. Polakowski, C. Adelmann, M. I. Popovici, S. V. Kalinin, T. Mikolajick, Impact of different dopants on the switching properties of ferroelectric hafniumoxide, *Jpn. J. Appl. Phys.*, 53, 08LE02 (2014).

16) D. Zhou, J. Xu, Q. Li, Y. Guan, F. Cao, X. Dong, J. Müller, **T. Schenk**, U. Schröder, Wake-up effects in Si-doped hafnium oxide ferroelectric thin films, *Appl. Phys. Lett.*, 103, 192904 (2013).

17) E.-C. Chang, D. Mikolas, P.-T. Lin, **T. Schenk**, C.-L. Wu, C.-K. Sung, C.-C. Fu, Improving feature size uniformity from interference lithography systems with non-uniform intensity profiles, *Nanotechnology*, 24, 455301 (2013).

www.ingramcontent.com/pod-product-compliance
Lightning Source LLC
Chambersburg PA
CBHW071548240526
45470CB00022B/351